Overcoming Data Scarcity in Earth Science

Overcoming Data Scarcity in Earth Science

Special Issue Editors

Angela Gorgoglione
Alberto Castro Casales
Christian Chreties Ceriani
Lorena Etcheverry Venturini

MDPI • Basel • Beijing • Wuhan • Barcelona • Belgrade

Special Issue Editors

Angela Gorgoglione
Universidad de la República
Uruguay

Alberto Castro Casales
Universidad de la República
Uruguay

Christian Chreties Ceriani
Universidad de la República
Uruguay

Lorena Etcheverry Venturini
Universidad de la República
Uruguay

Editorial Office
MDPI
St. Alban-Anlage 66
4052 Basel, Switzerland

This is a reprint of articles from the Special Issue published online in the open access journal *Data* (ISSN 2306-5729) from 2018 to 2020 (available at: https://www.mdpi.com/journal/data/special_issues/Data_Scarcity)

For citation purposes, cite each article independently as indicated on the article page online and as indicated below:

LastName, A.A.; LastName, B.B.; LastName, C.C. Article Title. *Journal Name* **Year**, *Article Number*, Page Range.

ISBN 978-3-03928-210-4 (Pbk)
ISBN 978-3-03928-211-1 (PDF)

Cover image courtesy of Chait Goli.

Contents

About the Special Issue Editors

Angela Gorgoglione received her Ph.D. in Civil and Environmental Engineering from Politecnico di Bari (Interpolytechnic Doctoral School—Politecnico di Bari, Milano, Torino) in 2016. She is currently an Assistant Professor at Universidad de la República, Uruguay. Her research applies hydraulic/hydrologic principles to improve the understanding of natural and urban systems and to contribute to solving significant environmental problems. Her research interests include water-quality modeling, hydrologic modeling, urban hydrology, and stormwater pollution.

Alberto Castro received his Ph.D. in Computer Architecture, major in Computer Networks at Universitat Politècnica de Catalunya, Spain, in 2014. He is currently an Assistant Professor at Universidad de la República, Uruguay. His research interests include communication networks, cognitive networks, and machine learning.

Christian Chreties received his Ph.D. in Engineering—Applied Fluid Mechanics at the School of Engineering, Universidad de la República, Uruguay. He is currently the Head of the Department of Fluid Mechanics and Environmental Engineering at the Universidad de la República, Uruguay, and has been an Associate Professor since 2004. His research work includes applied surface hydrology, fluvial hydraulics and sediment transport, and water resources management.

Lorena Etcheverry received her BE in Computer Engineering (2003), M.Sc. degree in Computer Science (2010), and Ph.D. in Computer Science (2016) from Universidad de la Republica, Uruguay. During her Ph.D., she worked at the Laboratory for Web & Information Technologies at Université Libre de Bruxelles (ULB), Belgium, and also at Instituto Tecnológico de Buenos Aires, Argentina. Since 2003, she has been with Universidad de la República, where she is currently an Assistant Professor. Her research interests are in the field of data management, in particular big data management, graph databases, data anonymization, and Semantic Web.

 *data*

Editorial

Overcoming Data Scarcity in Earth Science

Angela Gorgoglione [1],*, Alberto Castro [2], Christian Chreties [1] and Lorena Etcheverry [2]

[1] Department of Fluid Mechanics and Environmental Engineering (IMFIA), School of Engineering, Universidad de la República, Montevideo 11300, Uruguay; chreties@fing.edu.uy

[2] Department of Computer Science (InCo), School of Engineering, Universidad de la República, Montevideo 11300, Uruguay; acastro@fing.edu.uy (A.C.); lorenae@fing.edu.uy (L.E.)

* Correspondence: agorgoglione@fing.edu.uy

Received: 26 December 2019; Accepted: 30 December 2019; Published: 1 January 2020

Abstract: The Data Scarcity problem is repeatedly encountered in environmental research. This may induce an inadequate representation of the response's complexity in any environmental system to any input/change (natural and human-induced). In such a case, before getting engaged with new expensive studies to gather and analyze additional data, it is reasonable first to understand what enhancement in estimates of system performance would result if all the available data could be well exploited. The purpose of this Special Issue, "Overcoming Data Scarcity in Earth Science" in the *Data* journal, is to draw attention to the body of knowledge that leads at improving the capacity of exploiting the available data to better represent, understand, predict, and manage the behavior of environmental systems at meaningful space-time scales. This Special Issue contains six publications (three research articles, one review, and two data descriptors) covering a wide range of environmental fields: geophysics, meteorology/climatology, ecology, water quality, and hydrology.

Keywords: earth-science data; data scarcity; missing data; data quality; data imputation; statistical methods; machine learning; environmental modeling; environmental observations

---

1. Introduction

Environmental modeling deals with the representation of processes that occur in the real world in space and time. Based on differential equations, dynamic models mostly describe the processes that transform the environment through time. The spatial interactions and topological rules are mostly managed by geographic information systems (GIS) [1]. These mathematical models heavily rely on the data collected by direct field observations. However, a functional and complete dataset of any environmental variable is difficult to collect because of two main reasons: (i) the low reliability in the measurements (e.g., due to issues related to the equipment location or occurrences of equipment malfunctions); and (ii) the high cost of the monitoring campaigns [2,3]. The lack of an adequate amount of Earth-science data may induce an unsatisfactory and not reliable representation of the response's complexity of an environmental system to any input/change, both natural and human-induced. In this case, before undertaking expensive studies to collect and analyze additional environmental data, it is reasonable to first understand what improvement in estimates of system performance would result if all the available data could be well exploited [4].

Missing data imputation is a crucial task in cases where it is fundamental to use all available data and not neglect records with missing values [5]. Since the 1980s, many techniques to impute missing data have been proposed [6,7]. Generally speaking, the methods for filling in an incomplete dataset can be divided into two main categories: single imputation and multiple imputations [6]. Single imputation, i.e., filling in precisely one value for each missing one, intuitively has many appealing features, e.g., standard complete-data methods can be applied directly, and the substantial effort required to create imputations needs to be carried out only once. Multiple-imputation is a method of generating multiple simulated values for each missing item to reflect appropriately the uncertainty related to missing data [8].

A well-known and computationally simple method for the imputation of missing data is the mean substitution. However, it can disrupt the inherent structure of the data considerably, leading to significant errors in the covariance/correlation matrix and thereby degrading the performance of the model based on this data set [9]. A slightly better approach is to impute the missing elements from an ANOVA model [8]. More advanced imputation methods have been developed, and several methods and algorithms are now available.

The purpose of this Editorial is twofold: (i) combine and address the contributions of this Special Issue to use them as a basis in this area of science; (ii) encourage communication among the various disciplines by identifying and grouping complementary research solutions.

2. Summary

The main goal of the Special Issue "Overcoming Data Scarcity in Earth Science" in the *Data* journal, was to emphasize the body of knowledge that aims at enhancing the capacity of exploiting the available data to better characterize, understand, predict, and manage the behavior of environmental systems at all practical scales. This Special Issue contains six publications (three research articles, one review, and two data descriptors) covering a wide range of environmental disciplines: hydrology [10], water quality [11], meteorology/climatology [12,13], ecology [14], and geophysics [15].

2.1. Hydrology

In their article, Abraham et al. presented an application of machine learning for classifying soil into hydrologic groups [10]. Based on several soil characteristics such as the value of saturated hydraulic conductivity, and percentages of sand, silt, and clay, the authors trained machine learning models to classify soil into four hydrologic groups (Group A: soils with high infiltration rate and low runoff; Group B: soils with a moderate infiltration rate; Group C: soils with a slow infiltration rate; Group D: a very slow infiltration rate and high runoff potential). Afterward, they compared the results of the classification obtained using four different algorithms, (i) k-Nearest Neighbors (kNN), (ii) Support Vector Machine (SVM) with Gaussian Kernel, (iii) Decision Trees, (iv) Classification Bagged Ensembles and TreeBagger (Random Forest), with those obtained using estimation based on soil texture. Overall, kNN, Decision Tree, and TreeBagger performed better then SVM-Gaussian Kernel and Classification Bagged Ensemble. Among the four hydrologic groups, the authors noticed that group B had the highest rate of false positives.

2.2. Water Quality

Zavareh and Maggioni proposed an approach to analyzing water quality data based on rough set theory (RST) [11]. They collected six water quality indicators (temperature, pH, dissolved oxygen, turbidity, specific conductivity, and nitrate concentration) at the outlet of the catchment that contains the George Mason University campus in Fairfax (VA, United States) over three years (October 2015–December 2017). They evaluated the efficiency of using RST to estimate one water quality indicator based on other given (known) indicators. The authors stated that RST does not require any prior information on the dataset and represents a powerful tool able to deal with uncertainty and vagueness in the sample. Overall, RST was proven capable of finding primary indicators and discovering decision-making rules. RST-based decision-making rules can be a remarkable aid for analysts and planners for their decision-making process.

2.3. Meteorology/Climatology

In their work, Cazes Boezio and Ortelli evaluated the use of data-assimilation techniques from field measurements into initial conditions of atmospheric numerical simulations to obtain wind estimates in Uruguay (South America), at heights of 100 m above the ground and lower [12]. The wind was assessed with hourly frequency in a regular grid that covers the entire country. The field data to be assimilated was measured with anemometers placed 100 m above the ground in local wind farms. The data was assimilated into initial conditions for the Weather Research and Forecast regional model (WRF) of the National Center of Atmospheric Research (NCAR) using the module for data assimilation included in this model, the WRF-DA module. The authors stated that in addition to its direct use in the numerical prediction process, the results of data assimilation can be considered as "pseudo-observations" of atmospheric variables in regular grids.

In his data-descriptor publication, Mistry introduced a new high-resolution global gridded dataset of climate-extreme indices (CEIs) based on sub-daily precipitation and temperature data from the Global Land Data Assimilation System (GLDAS) [13]. This dataset, called "CEI_0p25_1970_2016", includes 71 annual (monthly in some cases) CEIs at $0.25° \times 0.25°$ gridded resolution, covering 47 years over the period 1970–2016. The author stated that CEI_0p25_1970_2016 fills gaps in existing CEI datasets by encompassing more indices and by being the only comprehensive global gridded CEI data available at high spatial resolution. The data of individual indices are freely downloadable in the commonly used Network Common Data Form 4 (NetCDF4) format. Potential applications of CEI_0p25_1970_2016 include the evaluation of sectoral impacts (e.g., hydrology, agriculture, energy, health), as well as the identification of spatial and temporal patterns that show similar historical of high/low temperature and precipitation extremes.

2.4. Ecology

In their thorough review, Pascoe et al. identified and discussed how the currently available environmental Earth data are lacking concerning their applications in species distribution modeling, mainly when predicting the potential distribution of invasive arthropods that vector pathogens (IAVPs) at significant space-time scales [14]. The authors examined the issues related to the interpolation of weather-station data, and the lack of microclimatic data, which is significant to the environment experienced by IAVPs. Furthermore, they provided some suggestions for filling these data gaps. The optimal resolution of environmental data relevant to IAVP ecology will likely vary according to the species under consideration, but they assumed that this resolution would typically be less than 1 m and hourly. The authors encourage modelers and ecologists to take a proactive approach in collecting small resolution data using data loggers, crowdsourcing, unmanned aerial vehicles or controlled environmental studies. They proposed that these proximally-sensed data, as well as remotely-sensed data, be made open access in a user-friendly database.

2.5. Geophysics

In their work, Bataleva et al. developed a sophisticated geophysical station that collects, processes, and store geophysical information, in particular, electrical and magnetic components of the natural electromagnetic field, useful for the study of geodynamic processes occurring in the Earth's crust and upper mantle [15]. This station is located in the territory of the Bishkek Geodynamic Proving Ground, located in the active seismic zone of the Northern Tien Shan (on the border between China and Kyrgyzstan, Central Asia).

3. Statistics

The following tables (from Tables 1–4) represent some statistics about the publications belonging to the Special Issue "Overcoming Data Scarcity in Earth Science" in the *Data* journal.

Table 1. Brief report of the Special Issue.

Submission	Quantity
Received	9
Published after review	6
Rejected	3
Acceptance rate	66.67%
Median publication time	57 days

Table 2. Type of publications belonging to the Special Issue.

Type of Publication	Quantity	Percentage
Article	3	50
Review	1	17
Data descriptor	2	33
Total	6	100

Table 3. Disciplines covered by the publications of the Special Issue.

Discipline	Quantity	Percentage
Hydrology	1	17
Water quality	1	17
Meteorology/climatology	2	33
Ecology	1	17
Geodynamics	1	17
Total	6	100

Table 4. Countries of the authors.

Country	Quantity	Percentage
Czech Republic	1	5
Italy	5	26
Kyrgyzstan	3	16
Netherland	1	5
United States	7	37
Uruguay	2	11
Total	18	100

Author Contributions: Conceptualization, A.G.; writing—original draft preparation, A.G.; writing—review and editing, A.C., C.C., and L.E. All authors have read and agreed to the published version of the manuscript.

Funding: This research received no external funding.

Acknowledgments: We gratefully acknowledge the technical and administrative support of the *Data* journal team. We also want to thank the Authors who contributed towards this Special Issue on "Overcoming Data Scarcity in Earth Science", as well as the Reviewers who provided the authors with suggestions and constructive feedback.

Conflicts of Interest: The authors declare no conflict of interest.

References

1. Chaulya, S.K.; Prasad, G.M. Chapter 7—Application of cloud computing technology in mining industry. In *Sensing and Monitoring Technologies for Mines and Hazardous Areas*; Elsevier: Amsterdam, The Netherlands, 2016; pp. 351–396.
2. Gorgoglione, A.; Bombardelli, F.A.; Pitton, B.J.L.; Oki, L.R.; Haver, D.L.; Young, T.M. Uncertainty in the parameterization of sediment build-up and wash-off processes in the simulation of water quality in urban areas. *Environ. Model. Softw.* **2019**, *111*, 170–181. [CrossRef]

3. Gorgoglione, A.; Gioia, A.; Iacobellis, V.; Piccinni, A.F.; Ranieri, E. A rationale for pollutograph evaluation in ungauged areas, using daily rainfall patterns: Case studies of the Apulian region in Southern Italy. *Appl. Environ. Soil Sci.* **2016**, *2016*, 9327614. [CrossRef]

4. Gorgoglione, A.; Gioia, A.; Iacobellis, V. A Framework for assessing modeling performance and effects of rainfall-catchment-drainage characteristics on nutrient urban runoff in poorly gauged watersheds. *Sustainability* **2019**, *11*, 4933. [CrossRef]

5. Jerez, J.M.; Molina, I.; García-Laencina, P.J.; Alba, E.; Ribelles, N.; Martín, M.; Franco, L. Missing data imputation using statistical and machine learning methods in a realbreast cancer problem. *Artif. Intell. Med.* **2010**, *50*, 105–115. [CrossRef] [PubMed]

6. Little, R.J.; Rubin, D.B. *Statistical Analysis with Missing Data*; John Wiley & Sons: Hoboken, NJ, USA, 2002.

7. Schafer, J.L. *Analysis of Incomplete Multivariate Data*; CRC Press: Boca Raton, FL, USA, 2010.

8. Junninen, H.; Niska, H.; Tuppurainen, K.; Ruuskanen, J.; Kolehmainen, M. Methods for imputation of missing values in air quality data sets. *Atmosph. Environ.* **2004**, *38*, 2895–2907. [CrossRef]

9. Tutz, G.; Ramzan, S. Improved methods for the imputation of missing data by nearest neighbor methods. *Comput. Stat. Data Anal.* **2015**, *90*, 84–99. [CrossRef]

10. Abraham, S.; Huynh, C.; Vu, H. Classification of soils into hydrologic groups using machine learning. *Data* **2020**, *5*, 2. [CrossRef]

11. Zavareh, M.; Maggioni, V. Application of rough set theory to water quality analysis: A case study. *Data* **2018**, *3*, 50. [CrossRef]

12. Cazes Boezio, G.; Ortelli, S. Use of the WRF-DA 3D-Var data assimilation system to obtain wind speed estimates in regular grids from measurements at wind farms in Uruguay. *Data* **2019**, *4*, 142. [CrossRef]

13. Mistry, M.N. A high-resolution global gridded historical dataset of climate extreme indices. *Data* **2019**, *4*, 41. [CrossRef]

14. Pascoe, E.L.; Pareeth, S.; Rocchini, D.; Marcantonio, M. A Lack of "environmental earth data" at the microhabitat scale impacts efforts to control invasive arthropods that vector pathogens. *Data* **2019**, *4*, 133. [CrossRef]

15. Bataleva, E.; Rybin, A.; Matiukov, V. System for collecting, processing, visualization, and storage of the MT-Monitoring data. *Data* **2019**, *4*, 99. [CrossRef]

 *data*

Article

Classification of Soils into Hydrologic Groups Using Machine Learning

Shiny Abraham *, Chau Huynh and Huy Vu

Department of Electrical and Computer Engineering, Seattle University, Seattle, WA 98122, USA;
huynhc3@seattleu.edu (C.H.); vuh8@seattleu.edu (H.V.)
* Correspondence: abrahash@seattleu.edu

Received: 1 October 2019; Accepted: 15 December 2019; Published: 19 December 2019

Abstract: Hydrologic soil groups play an important role in the determination of surface runoff, which, in turn, is crucial for soil and water conservation efforts. Traditionally, placement of soil into appropriate hydrologic groups is based on the judgement of soil scientists, primarily relying on their interpretation of guidelines published by regional or national agencies. As a result, large-scale mapping of hydrologic soil groups results in widespread inconsistencies and inaccuracies. This paper presents an application of machine learning for classification of soil into hydrologic groups. Based on features such as percentages of sand, silt and clay, and the value of saturated hydraulic conductivity, machine learning models were trained to classify soil into four hydrologic groups. The results of the classification obtained using algorithms such as k-Nearest Neighbors, Support Vector Machine with Gaussian Kernel, Decision Trees, Classification Bagged Ensembles and TreeBagger (Random Forest) were compared to those obtained using estimation based on soil texture. The performance of these models was compared and evaluated using per-class metrics and micro- and macro-averages. Overall, performance metrics related to kNN, Decision Tree and TreeBagger exceeded those for SVM-Gaussian Kernel and Classification Bagged Ensemble. Among the four hydrologic groups, it was noticed that group B had the highest rate of false positives.

Keywords: multi-class classification; soil texture calculator; k-Nearest Neighbors; support vector machines; decision trees; ensemble learning

---

1. Introduction

Soils play a crucial role in the global hydrologic cycle by governing the rates of infiltration and transmission of rainfall, and surface runoff, i.e., precipitation that does not infiltrate into the soil and runs across the land surface into water bodies, such as streams, rivers and lakes. Runoff occurs when rainfall exceeds the infiltration capacity of soils, and it is based on the physical nature of soils, land cover, hillslope, vegetation and storm properties such as rainfall duration, amount and intensity. The rainfall-runoff process serves as a catalyst for the transport of sediments and contaminants, such as fertilizers, pesticides, chemicals and organic matter, negatively impacting the morphology and biodiversity of receiving water bodies [1,2]. Flooding and erosion caused by uncontrolled runoff, particularly downstream, results in damage to agricultural lands and manmade structures [1]. Hence, modeling surface runoff is an essential part of soil and water conservation efforts, including but not limited to, forecasting floods and soil erosion and monitoring water and soil quality.

The U.S. Department of Agriculture's (USDA) agency for Natural Resources Conservation Service (NRCS), formerly known as the Soil Conservation Service (SCS), developed a parameter called Curve Number (CN) to estimate the amount of surface runoff. Furthermore, soils are classified into Hydrologic Soil Groups (HSGs) based on surface conditions (infiltration rate) and soil profiles (transmission rate). Combinations of HSGs and land use and treatment classes form hydrologic soil-cover complexes, each of which is assigned a CN [3]. A higher CN indicates a higher runoff potential. Consequently, accurate classification of HSGs is critical for the calculation of CNs that provide a meaningful prediction of runoff.

In the United States, more than 19,000 soil series have been identified and aggregated into map unit components with similar physical and runoff characteristics, and assigned to one of four HSGs: A, B, C or D. The original assignments were based on measured rainfall, runoff and infiltrometer data [4]. Since then, assignments have been based on the judgement of soil scientists, primarily relying on their interpretation of criteria published in the National Engineering Handbook (NEH) Part 630, Hydrology [5]. As with any subjective interpretation, the placement of soils into appropriate hydrologic groups have been non-uniform and inconsistent over time and across geographical locations. Soils with similar runoff characteristics were placed in the same hydrologic group, under the assumption that soils found within a climatic region with similar depth, permeability and texture will have similar runoff responses. Conventional soil mapping techniques extrapolate these classifications and geo-reference them with GPS (Global Positioning Systems) and digital elevation models visualized in a GIS (Geographic Information Systems) [6,7]. However, in addition to the inconsistent classification of soil profiles, the varying definition of mapping units introduces a certain degree of subjectivity. Over the past two decades, Pedology research has witnessed an evolution from traditional soil mapping techniques to methods for 'the creation and population of spatial soil information systems by numerical models inferring the spatial and temporal variations or soil types and soil properties from soil observation and knowledge and from related environmental variables' [8], also known as Digital Soil Mapping (DSM) [9–11].

Considering the advances in modern computing and the vastly expanding soil databases, NRCS and the Agricultural Research Service (ARS) formed a joint working group in 1990 to address shortcomings attributed to guidelines stated in NEH reference documents [12]. Two among the several goals identified by the group were to standardize the procedure for the calculation of CNs from rainfall-runoff data and to reconsider the HSG classifications. A fuzzy model that was developed using the National Soil Information System (NASIS) soil interpretation subsystem was applied to 1828 unique soils using data from Kansas, South Dakota, Missouri, Iowa, Wyoming and Colorado. Correlation between the soil's assigned and modeled HSG was analyzed, and the overall HSG frequency coincidence exceeded 54 percent [13]. It was observed that the correlation frequencies for soils from groups A and D were higher than those for groups B and C. These correlation inconsistencies were attributed to: (1) boundary conditions that occur when soils exhibit properties that do not fit entirely into a single hydrologic group. The effects of this are more profound for groups B and C considering that they are each bounded by two groups (2) fuzzy modeling of the subjective HSG criteria. To address the inconsistencies due to boundary conditions, an improved method that developed an automated system based on detailed soil attribute data was proposed by Li, R et al. [14]. This work aimed to mitigate the aggregation effect of HSGs on soil information, and eventually the CNs, due to the assignment of similar soils into different HSGs (exaggerating small differences between them) or different soils to the same HSG (omitting differences between them). Furthermore, this work successfully identified improper placement of HSGs. However, this work used a significantly smaller sample size of 67 soil types in the Lake Fork watershed in Texas.

Machine learning, a branch of Artificial Intelligence, is an inherently interdisciplinary field that is built on concepts such as probability and statistics, information theory, game theory and optimization, among many others. In 1959, Arthur Samuel, one of the pioneers of machine learning, defined machine learning as a "field of study that gives computers the ability to learn without being explicitly programmed" [15]. A more recent and widely accepted definition can be attributed to Tom Mitchell: "A computer program is said to learn from experience E with respect to some class of tasks T and performance measure P, if its performance at tasks in T, as measured by P, improves with experience E" [16]. Based on the approach used, type of input and output data, and nature of the problem being addressed, machine learning techniques can be classified into four main categories: (1) supervised learning; (2) unsupervised learning; (3) semi-supervised learning; and (4) reinforcement learning.

In supervised learning, the goal is to infer a function or mapping from training data that is labeled. The training data consist of an input vector X and an output vector Y that is labeled based on available prior experience. Regression and classification are two categories of algorithms that are based on supervised learning. Unsupervised learning, on the other hand, deals with unlabeled data, with the goal of finding a hidden structure or pattern in this data. Clustering is one of the most widely used unsupervised learning methods. In semi-supervised learning, a combination of labeled and unlabeled data is used to generate an appropriate model for the classification of data. The reinforcement learning method uses observations gathered from the interaction with the environment to make a sequence of decisions that would maximize the reward or minimize the risk. Q-learning is an example of a reinforcement learning algorithm.

The application of machine learning techniques in soil sciences ranges from the prediction of soil classes using DSM [17,18] to the classification of sub-soil layers using segmentation and feature extraction [19]. The predictive ability of machine learning models has been leveraged for agricultural planning and mass crop yield, the prediction of natural hazards, including, but not limited to, landslides, floods, drought and forest fires and monitoring the effects of climate change on the physical and chemical properties of soil [20,21]. Based on high spatial resolution satellite data, terrain/climatic data, and laboratory soil samples, the spatial distribution of six soil properties including sand, silt, and clay were mapped in an agricultural watershed in West Africa [22]. Of the four statistical prediction models tested and compared, i.e., Multiple Linear Regression (MLR), Random Forest Regression (RFR), Support Vector Machine (SVM) and Stochastic Gradient Boosting (SGB), machine learning algorithms performed generally better than MLR for the prediction of soil properties at unsampled locations. In a similar study for a steep-slope watershed in southeastern Brazil [23], the performance of three algorithms: Multinomial Logistic Regression (MLR), C5-decision tree (C5-DT) and Random Forest (RF) was evaluated and compared based on performance metrices of overall accuracy, standard error, and kappa index. It was observed that the RF model consistently outperformed the other models, while the MLR model had the lowest overall accuracy and kappa index. In the context of DSM applications, complex models such as RF are found to be better classifiers than generalized linear models such as MLR. While machine learning offers the added advantage of identifying trends and patterns with continuous improvement over time, these models are only as good as the quality of the data collected. An unbiased and inclusive dataset, along with the right choice of model, parameters, cross-validation method, and performance metrices is necessary to achieve meaningful results.

In this work, we investigated the application of four machine learning methods: kNN, SVM-Gaussian Kernel, Decision Trees and Ensemble Learning towards the classification of soil into hydrologic groups. The results of these algorithms are compared to those obtained using estimation based on soil texture.

2. Background

Soils are composed of mineral solids derived from geologic weathering, organic matter solids consisting of plant or animal residue in various stages of decomposition, and air and water that fill the pore space when soil is dry and wet, respectively. The mineral solid fraction of soil is composed of sand, silt and clay, relative percentages of which determine the soil texture in accordance with the USDA system of particle-size classification. Sand, being the larger of the three, feels gritty, and ranges in size from 0.05 to 2.00 mm. Sandy soils have poor water-holding capacity that can result in leaching loss of nutrients. Silt, being moderate in size, has a smooth or floury texture, and ranges from 0.002 to 0.05 mm. Clay, being the smallest of the three, feels sticky, and is made up of particles smaller than 0.002 mm in diameter. In general, the higher the percentage of silt and clay particles in soil, the higher is its water-holding capacity. Particles larger than 2.0 mm are referred to as rock fragments and are not considered in determining soil texture, although they can influence both soil structure and soil–water relationships. The ease with which pores in a saturated soil transmit water is known as saturated hydraulic conductivity (Ksat), and it is expressed in terms of micrometers per second (or inches per hour). Pedotransfer functions (PTFs) are commonly used to estimate Ksat in terms of readily available soil properties such as particle size distribution, bulk density, and organic matter content [24,25]. Machine Learning-based PTFs have been developed to understand the relationship between soil hydraulic properties and soil physical variables [26].

Hydrologic Soil Groups

Soils are classified into HSGs based on the minimum rate of infiltration obtained for bare soil after prolonged wetting [5]. The four hydrologic soil groups (HSGs) are described as follows:

Group A—Soils in this group are characterized by low runoff potential and high infiltration rates when thoroughly wet. They typically have less than 10 percent clay and more than 90 percent sand or gravel. The saturated hydraulic conductivity of all soil layers exceeds 40.0 micrometers per second.

Group B—Soils in this group have moderately low runoff potential and moderate infiltration rates when thoroughly wet. They typically have between 10 and 20 percent clay and 50 to 90 percent sand. The saturated hydraulic conductivity ranges from 10.0 to 40.0 micrometers per second.

Group C—Soils in this group have moderately high runoff potential and low infiltration rates when thoroughly wet. They typically have between 20 and 40 percent clay and less than 50 percent sand. The saturated hydraulic conductivity ranges from 1.0 to 10.0 micrometers per second.

Group D—Soils in this group are characterized by high runoff potential and very low infiltration rates when thoroughly wet. They typically have greater than 40 percent clay and less than 50 percent sand. The saturated hydraulic conductivity is less than or equal to 1.0 micrometers per second.

Dual hydrologic soil groups—Certain wet soils are placed in group D based solely on the presence of a high water table. Once adequately drained, they are assigned to dual hydrologic soil groups (A/D, B/D and C/D) based on their saturated hydraulic conductivity. The first letter applies to the drained condition and the second to the undrained condition.

3. Methods

3.1. Soil Survey Data

The dataset used for this work was obtained from USDA's NRCS Web Soil Survey (WSS), the largest public-facing natural resource database in the world [27]. The Soil Survey Geographic Database (SSURGO) developed by the National Cooperative Soil Survey was used to identify Areas of Interests (AOI) in the State of Washington the Idaho Panhandle National Forest. Tabular data corresponding to Physical Soil Properties and Revised Universal Soil Loss Equation, Version 2 (RUSLE2) related attributes for various AOIs were retrieved from the Microsoft Access database and compiled into Microsoft Excel spreadsheets. Features of interest include the map symbol and soil name, its corresponding hydrologic group, percentages of sand, silt and clay, depth in inches and Ksat in micrometers per second. The initial dataset comprised of 4468 unique soil types.

As with most survey-based datasets, there were incomplete or missing data, inconsistencies in formatting and undesired data entries. The compiled dataset was preprocessed to remove samples corresponding to: missing data points, dual hydrologic groups (A/D, B/D and C/D), and soil layers beyond a water impermeable depth range of 20 to 40 inches. This reduced the dataset to 2107 unique soil types. MATLAB® programming environment was used for all data preparation and processing.

3.2. Estimation Based on Soil Texture

Based on the percentages of sand, silt, and clay, soils can be grouped into one of the four major textural classes: (1) sands; (2) silts; (3) loams; and (4) clays. The soil textural triangle shown in Figure 1 illustrates twelve textural classes as defined by the USDA [28]: sand, loamy sand, sandy loam, loam, silt loam, silt, sandy clay loam, clay loam, silty clay loam, sandy clay, silty clay, and clay. These classifications are typically named after the primary constituent particle size, e.g., "sand", or a combination of the most abundant particles sizes, e.g., "sandy clay". One side of the triangle represents percent sand, the second side represents percent clay, and the third side represents percent silt. Given the percentages of sand, silt and clay in the soil sample, the corresponding textural class can be read from the triangle. Alternately, the NRCS soil texture calculator [28] can be used to determine textural class based on specific relationships between sand, silt and clay percentages as shown in Table 1. In this work, the method used to assign HSGs based on soil texture was adopted from Hong and Adler (2008) [29], which was modified from the USDA handbook [30] and National Engineering Handbook Section 4 [5]. MATLAB® was used to assign HSGs based on soil texture calculations.

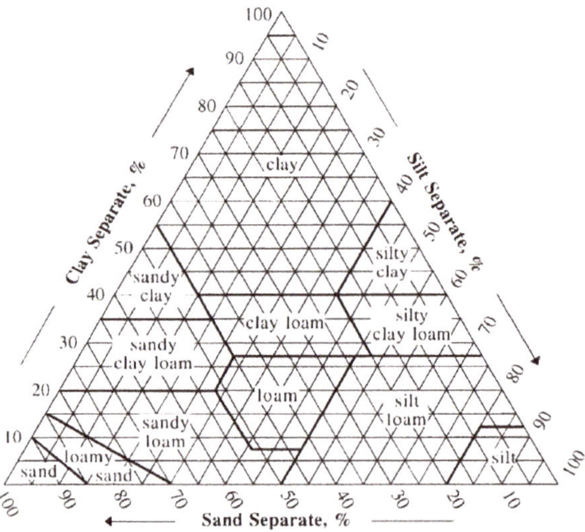

Figure 1. The soil textural triangle is used to determine soil textural class from the percentages of sand, silt and clay in the soil [28].

Table 1. Soil texture calculations and mapping to hydrologic soil groups [28,29].

Relationship between Sand, Silt and Clay Percentages	Textural Class	Hydrologic Soil Group
((silt + 1.5 * clay) < 15)	SAND	A
((silt + 1.5 * clay ≥ 15) AND (silt + 2 * clay < 30))	LOAMY SAND	A
((clay ≥ 7 && clay < 20) AND (sand > 52) AND ((silt + 2 * clay) ≥ 30) OR (clay < 7 && silt < 50 AND (silt + 2 * clay) ≥ 30))	SANDY LOAM	A
((clay ≥ 7 AND clay < 27) AND (silt ≥ 28 AND silt < 50) AND (sand ≤ 52))	LOAM	B
((silt ≥ 50 AND (clay ≥ 12 AND clay < 27)) OR ((silt ≥ 50 AND silt < 80) AND clay < 12))	SILT LOAM	B
(silt ≥ 80 AND clay < 12)	SILT	B
((clay ≥ 20 AND clay < 35) AND (silt < 28) AND (sand > 45))	SANDY CLAY LOAM	C
((clay ≥ 27 AND clay < 40) AND (sand > 20 AND sand ≤ 45))	CLAY LOAM	D
((clay ≥ 27 AND clay < 40) AND (sand ≤ 20))	SILTY CLAY LOAM	D
(clay ≥ 35 AND sand > 45)	SANDY CLAY	D
(clay ≥ 40 AND silt ≥ 40)	SILTY CLAY	D
clay ≥40 AND sand ≤ 45 AND silt < 40	CLAY	D

3.3. Machine Learning Algorithms

A common problem encountered in machine learning and data science is that of overfitting, where the model does not generalize well from training data to unseen data. Cross validation techniques are generally used to assess the generalization ability of a predictive model, thus avoiding the problem of overfitting. In this work, a Monte Carlo Cross-Validation (MCCV) method [31] was used by randomly splitting the dataset into equal-sized training and test subsets, training the model, predicting classification and repeating the process 100 times. The overall prediction accuracy (or other performance metrics) is the average over all iterations.

A machine learning algorithm can be classified as either parametric or non-parametric. Parametric methods assume a finite and fixed set of parameters, independent of the number of training examples. In non-parametric methods, also called instance-based or memory-based learning, the number of parameters is determined in part by the data, i.e., the number of parameters grows with the size of the training set. Due to the availability of a large dataset with labeled data, in this work, we considered four non-parametric supervised learning algorithms: (1) kNN (2) SVM Gaussian Kernel (3) Decision Trees (4) Random Forest. A qualitative introduction to these algorithms is presented in the following subsections.

3.3.1. k-Nearest Neighbors (kNN) Algorithm

kNN algorithm, an instance-based method of learning, is based on the principle that instances within a dataset will generally exist in close proximity to other instances that have similar properties. If the instances are tagged with a classification label, then the value of the label of an unclassified instance can be predicted based on the labels of its nearest neighbors.

The Statistics and Machine Learning Toolbox from MATLAB® was used to create a Classification kNN model using function *'fitcknn'*, followed by the function *'predict'* to predict classification for test data.

knn_model = *fitcknn* (features, labels, 'NumNeighbors', k)
predict_HSG = *predict* (knn_model, features_test)

where *features* is a numeric matrix that contains percent sand, percent silt, percent clay and Ksat; *label* is a cell array of character vectors that contain the corresponding HSGs; and k represents the number of neighbors.

3.3.2. Support Vector Machines (SVMs) with Gaussian Kernel

Support Vector Machines are non-parametric, supervised learning models that are motivated by a geometric idea of what makes a classifier "good" [32]. For linearly separable data points, the objective of the SVM algorithm is to find an optimal hyperplane (or a decision boundary) in an N-dimensional space (where N is the number of features) that distinctly classifies data points. Support vectors are the data points that lie closest to the hyperplane. The SVM algorithm aims to maximize the margin around the separating hyperplane, essentially making it a constrained optimization problem.

For data points that are not linearly separable, which is true of most real-world data, the features can be mapped into a higher-dimensional space in such a way that the classes become more easily separated than in the original feature space. A technique commonly referred to as the 'kernel trick', uses a kernel function that defines the inner product of the mapping functions in the transformed space. One of the most popular kernels are the Radial Basis Functions (RBFs), of which, the Gaussian kernel is a special case.

The Statistics and Machine Learning Toolbox from MATLAB® was used to create a template for SVM binary classification based on a Gaussian kernel function using function *'templateSVM'*, followed by the function *'fitcecoc'* that trains an Error-Correcting Output Codes (ECOC) model based on the features and labels provided. t is specified as a binary learner for an ECOC multiclass model. Finally, the function *'predict'* is used to predict classification for test data.

t = *templateSVM* ('KernelFunction', 'gaussian')
SVM_gaussian_model = *fitcecoc* (features, labels, 'Learners', t);
predict_HSG = *predict* (SVM_gaussian_model, features_test)

3.3.3. Decision Trees

Decision Trees are hierarchical models for supervised learning in which the learned function is represented by a decision tree [16,33]. The model classifies instances by querying them down the tree from the root to a leaf node, where each node represents a test over an attribute, each branch denotes its outcomes and each leaf node represents one class. Based on the measure used to select input variables and the type of splits at each node, decision trees can be implemented using statistical algorithms such as CART (Classification And Regression Tree), ID3 (Iterative Dichotomiser 3) and C4.5 (successor of ID3), among many others.

The Statistics and Machine Learning Toolbox from MATLAB® was used to grow a fitted binary classification decision tree based on the features and labels using function '*fitctree*', followed by the function '*predict*' to predict classification for test data. Function '*fitctree*' uses the standard CART algorithm to grow decision trees.

decisiontree_model = *fitctree* (features, labels);
predict_HSG = *predict* (decisiontree_model, features_test)

3.3.4. Ensemble Learning

While decision trees are a popular choice for predictive modeling due to their inherent simplicity and intuitiveness, they are often characterized by high variance. Consequently, decision trees can be unstable because small variations in the data might result in a completely different tree and hence, a different prediction. Ensemble learning methods that combine and average over multiple decision trees have been used to improve predictive performance [32]. Bagging (or bootstrap aggregation) is a technique that is used to generate new datasets with approximately the same (unknown) sampling distribution as any given dataset. Random forests, an extension of the bagging method, also selects a random subset of features. In other words, random forests can be considered as a combination of 'bootstrapping' and 'feature bagging'.

The Statistics and Machine Learning Toolbox from MATLAB® was used to grow an ensemble of learners for classification using function '*fitcensemble*'', followed by the function '*predict*' to predict classification for test data.

ensemble_model = *fitcensemble* (features, labels);
predict_HSG = *predict* (ensemble_model, features_test);

The function '*TreeBagger*' bags an ensemble of decision trees for classification using the Random Forest algorithm, followed by the function '*predict*' to predict classification for test data. Decision trees in the ensemble are grown using bootstrap samples of the data, with a random subset of features to use at each decision split.

treebagger_model = *TreeBagger* (50, features, labels, '*OOBPrediction*', '*On*', '*OOBPredictorImportance*', '*On*');
predicted_HSG = *predict* (treebagger_model, features_test);

'*OOBPrediction*' and '*OOBPredictorImportance*' are set to '*on*' to store information on what observations are out of bag for each tree and to store out-of-bag estimates of feature importance in the ensemble, respectively.

4. Performance Metrics

A Confusion matrix is commonly used to visualize the performance of a classification algorithm. Figure 2 illustrates the confusion matrix for a multi-class model with N classes [34]. Observations on correct and incorrect classifications are collected into the confusion matrix $C(c_{ij})$, where c_{ij} represents the frequency of class i being identified as class j. In general, the confusion matrix provides four types of classification results with respect to one classification target k:

- True Positive (TP)—correct prediction of the positive class ($c_{k,k}$)
- True Negative (TN)—correct prediction of the negative class ($\sum_{i,j\in N\backslash\{k\}} c_{ij}$)
- False Positive (FP)—incorrect prediction of the positive class ($\sum_{i\in N\backslash\{k\}} c_{ik}$)
- False Negative (FN)—incorrect prediction of the negative class ($\sum_{i\in N\backslash\{k\}} c_{ki}$)

Figure 2. Confusion matrix for a multi-class model with N classes [34].

Several performance metrics can be derived from these four outcomes. The ones of interest to us are listed below, for per-class classifications:

- Accuracy: This metric simply measures how often the classifier makes a correct prediction.

$$Overall\ Accuracy = \frac{\sum_{i-1}^{N} c_{i,i}}{\sum_{i=1}^{N} \sum_{j=1}^{N} c_{i,j}} \tag{1}$$

- Recall (Sensitivity or True Positive Rate): This metric denotes the classifier's ability to predict a correct class

$$Recall_{class} = \frac{TP_{class}}{TP_{class} + FN_{class}} \tag{2}$$

- Precision: This metric represents the classifier's certainty of correctly predicting a given class

$$Precision_{class} = \frac{TP_{class}}{TP_{class} + FP_{class}} \tag{3}$$

- False Positive Rate (FPR): This metric represents the number of incorrect positive predictions out of the total true negatives

$$FPR_{class} = \frac{FP_{class}}{FP_{class} + TN_{class}} \tag{4}$$

- True Negative Rate (TNR or Specificity): This metric represents the number of correct negative predictions out of the total true negatives

$$TNR_{class} = \frac{TN_{class}}{FP_{class} + TN_{class}} \tag{5}$$

- F1-Score: This metric is a harmonic mean of precision and recall. Although the F1-score is not as intuitive as accuracy, it is useful in measuring how precise and robust the classifier is.

$$F1 - Score_{class} = \frac{2 * TP_{class}}{2 * TP_{class} + FN_{class} + FP_{class}} \tag{6}$$

- Matthews Correlation Coefficient (MCC): For binary classification, MCC summarizes into a single value the confusion matrix. This is easily generalizable to multi-class problems as well.

$$MCC_{class} = \frac{TP_{class} * TN_{class} - FP_{class} * FN_{class}}{\sqrt{(TP_{class} + FP_{class}) * (TP_{class} + FN_{class}) * (FP_{class} + TN_{class}) * (FN_{class} + TN_{class})}} \tag{7}$$

- Cohen's Kappa (κ): This metric compares an Observed Accuracy with an Expected Accuracy (random chance)

$$\kappa_{class} = \frac{p_o - p_e}{1 - p_e} \tag{8}$$

where p_o represents the accuracy and p_e represents a factor that is based on normalized marginal probabilities.

For multi-class classification problems, averaging per-class metric results can provide an overall measure of the model's performance. There are two widely used averaging techniques: macro-averaging and micro-averaging.

- Macro-average: Macro-averaging reduces the multi-class predictions down to multiple sets of binary predictions. The desired metric for each of the binary cases are calculated and then averaged resulting in the macro-average for the metric over all classes. For example, the macro-average for Recall is calculated as shown below:

$$Recall_{macro} = \frac{\sum_{i=1}^{N} Recall_i}{N} \tag{9}$$

- Micro-average: Micro-averaging uses individual true positives, true negatives, false positives and false negatives from all classes to calculate the micro-average. For example, the micro-average for Recall is calculated as shown below:

$$Recall_{micro} = \frac{\sum_{i=1}^{N} TP_{class}}{\sum_{i=1}^{N} TP_{class} + \sum_{i=1}^{N} FN_{class}} \tag{10}$$

Macro-averaging assigns equal weight to each class, whereas micro-averaging assigns equal weight to each observation. Micro-averages provide a measure of effectiveness on classes with large observations, whereas macro-averages provide a measure of effectiveness on classes with small observations.

5. Results and Discussions

Following data preparation and pre-processing in MATLAB®, soil data samples were classified into one of the four hydrologic groups using soil texture calculations, followed by classifications using the following algorithms: (a) k-Nearest Neighbor (kNN), (b) SVM Gaussian Kernel, (c) Decision Tree,

(d) Classification Bagged Ensemble, and (e) TreeBagger. A Monte Carlo Cross-Validation (MCCV) method was used to avoid the problem of overfitting [31]. A measure of overall accuracy was first computed to compare all five algorithms with the soil texture-based classification. Table 2 shows that overall accuracy for the latter is significantly lower than those for the machine learning-based classification algorithms. In fact, none of the HSG C occurrences were correctly classified using soil texture calculations. The TreeBagger (Random Forest) algorithm has the highest overall accuracy of 84.70 percent, closely followed by the Decision Tree and kNN algorithms with 83.12 percent and 80.66 percent, respectively. Although applied for an entirely different dataset, the fuzzy system hydrologic grouping model [13] results in an overall correlation frequency of 60.5 percent for HSGs A, B, C and D, with higher correlation between assigned and modeled results for HSGs A and D.

Table 2. Comparison of overall accuracy.

Method	Overall Accuracy
Soil Texture Calculator	0.54
k-Nearest Neighbor (kNN)	0.81
SVM Gaussian Kernel	0.72
Decision Tree	0.83
Classification Bagged Ensemble	0.79
TreeBagger	0.85

For datasets in which the classes are not represented equally (also known as imbalanced classes), accuracy is typically not a good measure of performance. Out of the 2107 unique soil samples in the observed group, 337 belong to HSG A, 1142 to HSG B, 511 to HSG C and 117 to HSG D. Given that our dataset is relatively imbalanced, we further evaluate the performance of all five algorithms based on metrics of Recall, Precision, FPR, TNR, F1-Score, MCC and Kappa. It is important to account for chance agreement when dealing with highly imbalanced classes since a high classification accuracy could result from classifying all observations as the largest class [35,36]. Table 3 lists per-class results and macro- and micro-averages of these metrics for classification using kNN, SVM and Decision Trees. Table 4 presents the same for two Ensemble Learning algorithms. A graphical comparison of individual classes (HSGs) for each metric is shown in Figure 3.

Table 3. Comparison of performance metrics for classification using k-Nearest Neighbors (kNN), Support Vector Machine (SVM) and decision trees.

k-Nearest Neighbor (kNN); k = 4							
	Recall	Precision	FPR	TNR	F1 Score	MCC	Kappa
HSG A	0.84	0.86	0.03	0.97	0.84	0.82	0.69
HSG B	0.85	0.84	0.20	0.80	0.84	0.65	0.08
HSG C	0.72	0.73	0.09	0.91	0.72	0.63	0.56
HSG D	0.73	0.83	0.01	0.99	0.77	0.76	0.89
Macro Average	0.78	0.81	0.08	0.92	0.79	0.72	0.56
Micro Average	0.80	0.80	0.07	0.93	0.80	0.73	0.73
Support Vector Machines (SVM) Gaussian Kernel							
HSG A	0.90	0.79	0.05	0.95	0.84	0.81	0.67
HSG B	0.86	0.71	0.42	0.58	0.77	0.46	0.09
HSG C	0.35	0.65	0.06	0.94	0.45	0.36	0.66
HSG D	0.54	0.98	0.00	1.00	0.69	0.72	0.91
Macro Average	0.66	0.78	0.13	0.87	0.69	0.59	0.58
Micro Average	0.74	0.74	0.09	0.91	0.74	0.65	0.65
Decision Tree							
HSG A	0.88	0.84	0.03	0.97	0.86	0.83	0.68
HSG B	0.91	0.83	0.22	0.78	0.87	0.70	0.03
HSG C	0.67	0.82	0.05	0.95	0.74	0.67	0.59
HSG D	0.66	0.85	0.01	0.99	0.74	0.73	0.90
Macro Average	0.78	0.84	0.08	0.92	0.80	0.73	0.55
Micro Average	0.79	0.79	0.07	0.93	0.79	0.72	0.72

Table 4. Comparison of performance metrics for classification using ensemble learning algorithms.

	Recall	Precision	FPR	TNR	F1 Score	MCC	Kappa
Classification Bagged Ensemble							
HSG A	0.89	0.79	0.05	0.95	0.83	0.80	0.67
HSG B	0.91	0.77	0.32	0.68	0.83	0.61	0.04
HSG C	0.49	0.82	0.04	0.96	0.61	0.55	0.64
HSG D	0.59	0.96	0.00	1.00	0.73	0.74	0.91
Macro Average	**0.72**	**0.83**	**0.10**	**0.90**	**0.75**	**0.68**	**0.57**
Micro Average	**0.79**	**0.79**	**0.07**	**0.93**	**0.79**	**0.73**	**0.73**
TreeBagger; N = 50							
HSG A	0.89	0.85	0.03	0.97	0.87	0.85	0.68
HSG B	0.93	0.84	0.21	0.79	0.88	0.73	0.02
HSG C	0.69	0.86	0.04	0.96	0.76	0.71	0.59
HSG D	0.65	0.87	0.01	0.99	0.74	0.74	0.90
Macro Average	**0.79**	**0.86**	**0.07**	**0.93**	**0.81**	**0.76**	**0.55**
Micro Average	**0.78**	**0.78**	**0.08**	**0.92**	**0.78**	**0.70**	**0.70**

It should be noted that micro-averages for Recall, Precision and F1-Sscore are equal, as expected in multi-class classification problems. Moreover, micro-averages for MCC and Kappa are equal. It can be observed that for all five algorithms, the ability of the classifiers to correctly predict (Recall) HSGs A and B are relatively higher when compared to HSGs C and D. This is in line with results obtained for three-class and seven-class classification of soil types using Decision Trees and SVM in [19], wherein sandy soils had higher classification accuracy. On the other hand, the certainty with which the classifiers predict correct classes (Precision) is relatively higher for HSG D in our work. A comparison of macro- and micro-averages of F1-Scores among the five classifiers shows that kNN, Decision Tree and TreeBagger have scores close to 0.8, while SVM-Gaussian Kernel lags with a score close to 0.7. Among the four soil groups, HSG B has the highest rate of False Positives, with the highest being 57.8 percent for SVM-Gaussian Kernel and lowest being 19.83 percent for kNN. The fact that HSG B is the largest class in the dataset, and bordered by two other groups, explains the high FPR. A comparison of macro- and micro-averages of MCCs among the five classifiers shows comparable results (~0.72) for kNN, Decision Tree and TreeBagger. Yet again, SVM-Gaussian kernel has the lowest score (~0.6). The results of Cohen's Kappa coefficient for HSG B shows some discrepancy that is consistent across all five classifiers. This may be related to the corresponding high FPRs. Regardless, the micro-average Kappa value is consistent with that of MCC, possibly accounting for any class imbalance. An interesting observation is that the micro- and macro-averages of Kappa coefficients for all five classifiers are similar in value. The macro-averages range from 0.55 (RF and DT) to 0.58 (SVM) and micro-averages range from 0.65 (SVM) to 0.73 (kNN and CBE), all within the moderate to substantial agreement range [37]. This similarity is observed in studies related to machine learning techniques for DSM, suggesting that the quality and robustness of datasets is of greater importance than the classifier itself [38,39]. In the context of predicting soil map units on tropical hillslopes in Brazil, an RF model yielded an overall accuracy of 78.8 percent and a Kappa index of 0.76, while a Decision Tree model had an overall accuracy of 70.2 percent and a Kappa value of 0.66 [39]. In contrast, for classification based on soil taxonomic units in British Columbia, Canada, kNN and SVM resulted in the highest accuracy of 72 percent; however, models such as CART with bagging and RF were preferred due to the speed of parameterization and the interpretability of the results, while resulting in similar accuracies ranging from 65 to 70 percent [40].

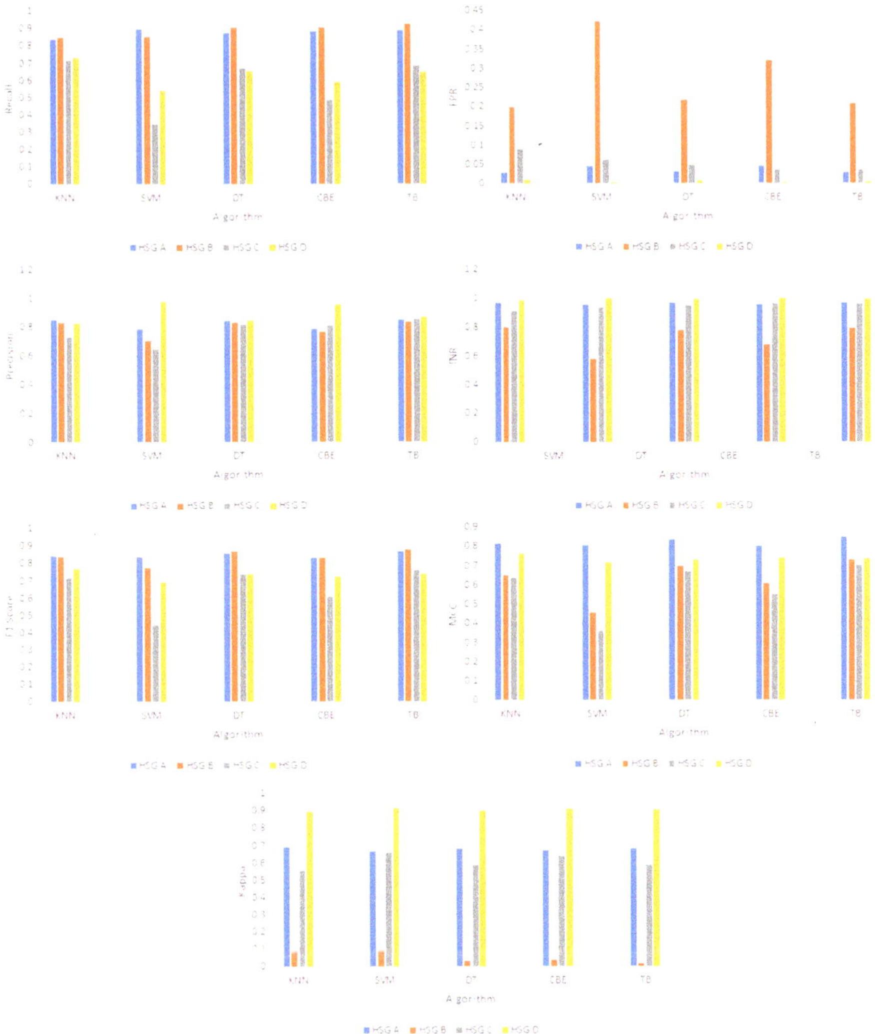

Figure 3. A graphical representation of per-class performance metrics for kNN, SVM-Gaussian Kernel, Decision Trees and Ensemble Learning Algorithms (CBE and TB).

6. Conclusions

This work presents the application of machine learning towards classification of soil into hydrologic groups. The machine learning models tested were kNN, SVM-Gaussian Kernel, Decision Trees and Ensemble Learning (Classification Bagged Ensemble and Random Forest). It was observed that for all five classifiers, Recall for HSGs A and B were relatively higher when compared to HSGs C and D, but precision was relatively higher for HSG D. Overall, performance metrics related to kNN, Decision Tree and TreeBagger exceeded those for SVM-Gaussian Kernel and Classification Bagged Ensemble.

As part of future work, the effects of class imbalance will be investigated by comparing datasets with varying degrees of imbalance and using various cross-validation techniques with proportional stratified random sampling. Deep learning methods that address this classification problem will also be explored.

Author Contributions: Conceptualization, S.A.; methodology, S.A.; software, S.A., C.H. and H.V.; validation, S.A., C.H. and H.V.; formal analysis, S.A., C.H. and H.V.; investigation, S.A., C.H. and H.V.; resources, S.A.; data curation, S.A., C.H. and H.V.; writing—original draft preparation, S.A.; writing—review and editing, S.A., C.H. and H.V.; visualization, S.A., C.H. and H.V.; supervision, S.A.; project administration, S.A.; funding acquisition, S.A. All authors have read and agreed to the published version of the manuscript.

Funding: This research was funded by W.M. Keck Foundation through the Undergraduate Education Grant Program.

Conflicts of Interest: The authors declare no conflict of interest.

References

1. Huffman, R.L.; Fangmeier, D.D.; Elliot, W.J.; Workman, S.R. Infiltration and Runoff. In *Soil and Water Conservation Engineering*, 7th ed.; American Society of Agricultural Engineers: St. Joseph, MI, USA, 2013; pp. 81–113.
2. Kokkonen, T.; Koivusalo, H.; Karvonen, T. A Semi-Distributed Approach to Rainfall-Runoff Modelling—A Case Study in a Snow Affected Catchment. *Environ. Model. Softw.* **2001**, *16*, 481–493. [CrossRef]
3. Hydrology Training Series: Module 104-Runoff Curve Number Computations. Available online: https://www.wcc.nrcs.usda.gov/ftpref/wntsc/H&H/training/runoff-curve-numbers1.pdf (accessed on 12 December 2018).
4. Musgrave, G.W. How Much of the Rain Enters the Soil? In *Water: U.S. Department of Agriculture Yearbook*; United States Government Publishing Office (GPO): Washington, DC, USA, 1995; pp. 151–159.
5. United States Department of Agriculture. Chapter 7: Hydrologic Soil Groups. In *Part 630 Hydrology, National Engineering Handbook*; 2009. Available online: https://directives.sc.egov.usda.gov/viewerFS.aspx?id=2572 (accessed on 10 December 2018).
6. Morris, D.K.; Stienhardt, G.C.; Nielsen, R.L.; Hostetter, W.; Haley, S.; Struben, G.R. Using GPS, GIS, and Remote Sensing as a Soil Mapping Tool. In Proceedings of the 5th International Conference on Precision Agriculture, Bloomington, IN, USA, 16–19 July 2000.
7. Usery, E.L.; Pocknee, S.; Boydell, B. Precision farming data management using geographic information systems. *Photogramm. Eng. Remote. Sens.* **1995**, *61*, 1383–1391.
8. Lagacherie, P.; McBratney, A.B.; Voltz, M. (Eds.) *Digital Soil Mapping: An Introductory Perspective*; Elsevier: Amsterdam, The Netherlands, 2007; Volume 31, pp. 3–24.
9. McBratney, A.B.; Mendonça-Santos, M.L.; Minasny, B. On digital soil mapping. *Geoderma* **2003**, *117*, 3–52. [CrossRef]
10. Carré, F.; McBratney, A.B.; Mayr, T.; Montanarella, L. Digital soil assessments: Beyond DSM. *Geoderma* **2007**, *142*, 69–79. [CrossRef]
11. Arrouyas, D.; McKenzie, N.; Hempel, J.; de Forges, A.R.; McBratney, A.B. *GlobalSoilMap: Basis of the Global Spatial Soil Information System*; Taylor and Francis: London, UK, 2014.
12. Runoff Curve Number Method: Beyond the Handbook. Available online: https://www.wcc.nrcs.usda.gov/ftpref/wntsc/H&H/CNarchive/CNbeyond.doc (accessed on 15 June 2019).
13. Neilsen, R.D.; Hjelmfelt, A.T. Hydrologic Soil-Group Assignment. In Proceedings of the International Water Resources Engineering Conference, Reston, VA, USA, 3–7 August 1998.
14. Li, R.; Rui, X.; Zhu, A.-X.; Liu, J.; Band, L.E.; Song, X. Increasing Detail of Distributed Runoff Modeling Using Fuzzy Logic in Curve Number. *Environ. Earth Sci.* **2015**, *73*, 3197–3205. [CrossRef]
15. Samuel, A.L. Some studies in machine learning using the game of checkers. *IBM J. research dev.* **1959**, *3*, 210–229. [CrossRef]
16. Mitchell, T.M. *Machine Learning*, 1st ed.; McGraw-Hill, Inc.: New York, NY, USA, 1997; p. 2.
17. Illés, G.; Kovács, G.; Heil, B. Comparing and Evaluating Digital Soil Mapping Methods in a Hungarian Forest Reserve. *Can. J. Soil Sci.* **2011**, *91*, 615–626. [CrossRef]
18. Behrens, T.; Schmidt, K.; MacMillan, R.A. Multi-Scale Digital Soil Mapping with Deep Learning. *Sci. Rep.* **2018**, *8*, 15244. [CrossRef]
19. Bhattacharya, B.; Solomatine, D. Machine learning in soil classification. *Neural Netw.* **2006**, *19*, 186–195. [CrossRef]

20. Tayfur, G.; Singh, V.P.; Moramarco, T.; Barbetta, S. Flood Hydrograph Prediction Using Machine Learning Methods. *Water* **2018**, *10*, 968. [CrossRef]

21. Yang, M.; Xu, D.; Chen, S.; Li, H.; Shi, Z. Evaluation of Machine Learning Approaches to Predict Soil Organic Matter and pH Using vis-NIR Spectra. *Sensors* **2019**, *19*, 263. [CrossRef] [PubMed]

22. Forkuor, G.; Hounkpatin, O.K.; Welp, G.; Thiel, M. High Resolution Mapping of Soil Properties Using Remote Sensing Variables in South-Western Burkina Faso: A Comparison of Machine Learning and Multiple Linear Regression Models. *PLoS ONE* **2017**, *12*, e0170478. [CrossRef] [PubMed]

23. Silva, B.P.C.; Silva, M.L.N.; Avalos, F.A.P.; Menezes, M.D.d.; Curi, N. Digital soil mapping including additional point sampling in Posses ecosystem services pilot watershed, southeastern Brazil. *Sci. Rep. Nat.* **2019**, *9*, 13763.

24. Wösten, J.H.M.; Pachepsky, Y.A.; Rawls, W.J. Pedotransfer functions: Bridging the gap between available basic soil data and missing soil hydraulic characteristics. *J. Hydrol.* **2001**, *251*, 123–150. [CrossRef]

25. Abdelbaki, A.M.; Youssef, M.A.; Naguib, E.M.F.; Kiwan, M.E.; El-giddawy, E.I. Evaluation of Pedotransfer Functions for Predicting Saturated Hydraulic Conductivity for U.S. Soils. In Proceedings of the American Society of Agricultural and Biological Engineers Annual International Meeting, Reno, NV, USA, 21–24 June 2009.

26. Araya, S.N.; Ghezzehei, T.A. Using machine learning for prediction of saturated hydraulic conductivity and its sensitivity to soil structural perturbations. *Water Resour. Res.* **2019**, *55*, 5715–5737. [CrossRef]

27. Natural Resources Conservation Service Web Soil Survey. Available online: https://websoilsurvey.sc.egov.usda.gov/App/HomePage.htm (accessed on 10 November 2018).

28. Natural Resources Conservation Service Soil Texture Calculator. Available online: https://www.nrcs.usda.gov/wps/portal/nrcs/detail/soils/survey/?cid=nrcs142p2_054167 (accessed on 10 November 2018).

29. Hong, Y.; Adler, R.F. Estimation of global SCS curve numbers using satellite remote sensing and geospatial data. *Int. J. Remote. Sens.* **2008**, *29*, 471–477. [CrossRef]

30. Urban Hydrology for Small Watersheds, Technical Release 55. Available online: www.nrcs.usda.gov/downloads/hydrology_hydraulics/tr55/tr55.pdf (accessed on 15 June 2019).

31. Xu, Q.S.; Liang, Y.Z. Monte Carlo cross validation. *Chemom. Intell. Lab. Syst.* **2001**, *56*, 1–11. [CrossRef]

32. Knox, S.W. *Machine Learning: A Concise Introduction*; John Wiley & Sons Inc.: Hoboken, NJ, USA, 2018.

33. Bell, J. *Machine Learning: Hands-On for Developers and Technical Professionals*; John Wiley & Sons Inc.: Hoboken, NJ, USA, 2014.

34. Kruger, F. Activity, Context, and Plan Recognition with Computational Causal Behavior Models. Ph.D. Thesis, University of Rostock, Mecklenburg, Germany, 2016. Available online: https://pdfs.semanticscholar.org/bebf/183d2f57f79b5b3e85014a9e1d6392ad0e5c.pdf (accessed on 10 June 2019).

35. Brungard, C.W.; Boettinger, J.L.; Duniway, M.C.; Wills, S.A.; Edwards, T.C., Jr. Machine learning for predicting soil classes in three semi-arid landscapes. *Geoderma* **2015**, *239*, 68–83. [CrossRef]

36. Congalton, R.G.; Green, K. *Assessing the Accuracy of Remotely Sensed Data: Principles and Practices*; CRC/Taylor & Francis: Boca Raton, FL, USA, 1998.

37. Landis, J.R.; Koch, G.G. The measurement of observer agreement for categorical data. *Biometrics* **1977**, *33*, 159–174. [CrossRef]

38. Meier, M.; de Souza, E.; Francelino, M.; Filho, E.I.F.; Schaefer, C.E.G.R. Digital Soil Mapping Using Machine Learning Algorithms in a Tropical Mountainous Area. *Revista Brasileira de Ciência do Solo* **2018**, *42*, 1–22. [CrossRef]

39. Chagas, C.S.; Pinheiro, H.; Carvalho, W.; Anjos, L.H.C.; Pereira, N.R.; Bhering, S.B. Data mining methods applied to map soil units on tropical hillslopes in Rio de Janeiro, Brazil. *Geoderma Reg.* **2016**, *9*, 47–55. [CrossRef]

40. Heung, B.; Ho, H.C.; Zhang, J.; Knudby, A.; Bulmer, C.; Schmidt, M. An overview and comparison of machine-learning techniques for classification purposes in digital soil mapping. *Geoderma* **2016**, *265*, 62–77. [CrossRef]

Article

Application of Rough Set Theory to Water Quality Analysis: A Case Study

Maryam Zavareh * and Viviana Maggioni

Department of Civil, Environmental and Infrastructure Engineering, George Mason University, Fairfax, VA 22030, USA; vmaggion@gmu.edu
* Correspondence: mzavareh@masonlive.gmu.edu; Tel.: +1-703-993-5117

Received: 27 September 2018; Accepted: 3 November 2018; Published: 7 November 2018

Abstract: This work proposes an approach to analyze water quality data that is based on rough set theory. Six major water quality indicators (temperature, pH, dissolved oxygen, turbidity, specific conductivity, and nitrate concentration) were collected at the outlet of the watershed that contains the George Mason University campus in Fairfax, VA during three years (October 2015–December 2017). Rough set theory is applied to monthly averages of the collected data to estimate one indicator (decision attribute) based on the remainder indicators and to determine what indicators (conditional attributes) are essential (core) to predict the missing indicator. The redundant attributes are identified, the importance degree of each attribute is quantified, and the certainty and coverage of any detected rule(s) is evaluated. Possible decision making rules are also assessed and the certainty coverage factor is calculated. Results show that the core water quality indicators for the Mason watershed during the study period are turbidity and specific conductivity. Particularly, if pH is chosen as a decision attribute, the importance degree of turbidity is higher than the one of conductivity. If the decision attribute is turbidity, the only indispensable attribute is specific conductivity and if specific conductivity is the decision attribute, the indispensable attribute beside turbidity is temperature.

Keywords: rough set theory; water quality; attribute reduction; core attribute; rule extraction

---

1. Introduction

Since water quality is affected by complex factors like animal/human activities and weather events, its continuous sampling and monitoring is of paramount importance for human health [1]. The United States Geological Survey (USGS) has been continuously monitoring the quality of surface water across the U.S. over the past decades [2]. The most common water quality indicators suggested by the USGS are temperature, specific conductance, dissolved oxygen concentration (DO), pH, and turbidity. Collecting and analyzing water quality data is a challenging task. First off, water quality monitoring techniques are different in different water bodies like streams, lakes, bays, and estuaries, characterized not only by different microscopic and macroscopic organisms, but also by different ecosystems, flow rate, and accessibility. Additional common challenges include uncertainty in water quality observations and instrument failure. In the instance of instrument malfunctioning or stop recording, one or more values in the time series may be missing. Popular methods to recover gaps in time series are divided into two major groups: deterministic and stochastic [3]. Examples of deterministic approaches are nearest-neighbor interpolation, polynomial interpolation, and methods based on distance weighting. Regression methods, auto regressive methods, and machine learning methods fall under the stochastic category [3].

Sampling water quality is further complicated by the development of an effective method to analyze and evaluate the collected data. Water quality data are usually characterized by non-Gaussian distributions. Also, the presence of outliers and missing values are very common [4]. As a result, finding an appropriate analytical method is key. Some popular classical methods are graphical analysis (e.g., boxplots, scatter plots, and Q-Q plots), probability distribution analysis, and trend analysis. However, when dealing with excessive amount of data, it is easy to miss hidden patterns and information. In the past two decades, several studies have proposed novel approaches to analyze water quality data, including fuzzy theory [5], maximum likelihood methods [6], principal component analysis [7], cascade correlation artificial neural network [8], interactive fuzzy multi-objective linear programming [9], linear regression [10], inexact chance-constrained quadratic programming [11], and Dempster-Shafer methods [12]. All these methods have the ability to deal with large datasets and investigate relationships among water quality indicators. However, to take advantage of the above tools, prior and/or additional information about the data is needed. For example, the fuzzy set theory requires a grade of membership (that defines how each data point is mapped to a membership value) or a value of possibility (e.g., possible, quite possible, slightly possible, impossible). Similarly, the Dempster-Shafer theory necessitates basic probability analysis [13].

Rough Set Theory (RST), introduced by Pawlak in 1982 [13], represents a valid alternative to overcome these issues. RST is a powerful tool to deal with large amounts of information, does not require preliminary or additional information about the data, and considers vagueness and uncertainty in the dataset [14]. RST is commonly used in classification, ranking, multi-criteria decision analysis, and decision rules [15]. One of the applications of RST is pattern recognition by attribute reduction. By reducing unnecessary features, RST is capable of discovering hidden patterns in high dimensional datasets [16]. The philosophy of rough set is based on the assumption that some information is associated to every object in the universe. Objects sharing the same information are called indiscernible and the indiscernibility relation is the mathematical basis of rough set theory [17]. This tool has been successfully applied to areas like healthcare, banking, medicine, engineering, environmental science, among others [17].

In this work, we investigate the potential of applying RST to water quality analysis. RST is useful when dealing with complexity and vagueness in a dataset, which is always the case when analyzing water quality field data. Although a few attempts exist in the field of environmental and water resources engineering [18,19], the application of RST for assessing water quality indicators has not been widely explored. For example, Shen and Chouchoulas [20] proposed a hybrid system called fuzzy-rough estimator to assess the size of algae population based on water characteristics. Although their attribute reduction method (going from eleven original attributes to seven) was demonstrated to be successful, their approach was not capable of extracting high accuracy sets of rules. Another application of RST in water resources engineering is the one investigated by Barbagallo et al. [21] who studied reservoir operating rules. This study employed the integrated RST and Rose application, a software developed by the University of Poznan in Poland [22], to provide the minimal condition attributes and reveal the relevance of each attribute. Dong et al. [18] proposed a model to forecast annual runoff from a reservoir using RST. Their results showed that the larger the samples was, the more accurate the model. In a study performed by Ip et al. [23], RST was employed to identify the significant water quality indicators in a decision-making system. Specifically, RST was able to reduce the number water quality indicators and quantify the importance degree of each core indicator.

Other studies combined RTS with other approaches, such as the one by Pai and Lee [19] that introduced the Multinomial Logistics Regression (MLR) model. MLR was used to investigate the relationship between different degrees of water pollution and environmental factors, like the one between the concentration of SO_2 emitted by car and motorcycle exhausts and ozone density in the atmosphere. This framework was shown to be capable of predicting water quality using environmental factors rather than monitoring the processes of chemical elements. Another example is the work by Karimi et al. [24] who employed the variable consistency dominance-based rough set approach to explore the complex relationship between water quality and environmental indicators. They explored the relationship between total dissolved solids (TDS) and environmental indicators used as explanatory variables, such as precipitation, river water temperature, runoff, normalized difference vegetation index (NVDI), land surface temperature, river water temperature. Using a moving average filter in the TDS data, they decreased the noise and reduced the width of the boundary region between the lower approximation (all elements in a subset belong to the set) and upper approximation (all elements possibly belong to the set).

The main goal of this work is to assess the efficiency of using RST to estimate one water quality indicator based on given (known) indicators. Evaluating the overall quality of the stream water is outside the scope of this work. What we consider here is a comprehensive approach that looks at several water quality indicators rather than providing a generic assessment of the stream healthiness. Our hypothesis is that, when observations in a time series are missing, RST is capable of providing information regarding the missing indicator based on the other recorded indicators. RST also identifies the dispensable indicators. By eliminating the dispensable (redundant) indicator or indicators, the complexity of the dataset is reduced. The strength of each indicator in finding an unknown indicator is assessed and dispensable attributes are identified to discover hidden patterns. Section 2 introduces the basics of rough set theory and its application to a water quality dataset collected in Fairfax, VA during 2015 to 2017. Section 3 presents the results, whereas Section 4 discusses the results and summarizes the main conclusions.

2. Materials and Methods

2.1. Rough Set Theory

In RST, data are represented by an information system or information table consisting of non-empty sets of finite objects (rows) and non-empty finite set of attribute (columns). More formally:

$$S = (U, A) \tag{1}$$

where S is the decision system, U is the universe, and A is an attribute.

The central concept in RST is the indiscernibility relation, a relationship between two (or more) objects where all the values are identical (equivalent) with respect to a subset of considered attributes [25]. The indiscernibility relation is defined as any subset B of A with a binary relation I (B) on U. For every $a \in A$: $(x, y) \in I(B)$ if and only if $a(x) = a(y)$, where the value of attribute a is for element x (or y).

Approximation is another fundamental concept in RST. On one hand, lower approximation refers to the domain of objects that are known with certainty to belong to the subset of interest. The lower approximation is also called B-positive region, $posB(X)$. On the other hand, upper approximation refers to objects that possibly belong to the subset of interest [26].

Suppose $X \subset U$, and $B \subset A$, the B_{lower} and B_{upper} approximation of X, respectively, are:

$$B_{lower}(X) = \cup \{B(x):B(x) \subset X\} \tag{2}$$

$$B_{upper}(X) = \cup \{B(x):B(x) \cap X \neq \varnothing\}. \tag{3}$$

Therefore, the B-boundary region of X is defined as:

$$BNB(X) = B_{upper}(X) - B_{lower}(X). \tag{4}$$

If the boundary region is empty, then X is exact (or crisp). Otherwise, X is inexact and is classified as rough. The approximation method is a valuable method to express data topological properties [14]. The decision-making (DM) rule is another helpful tool to discover hidden patterns in a dataset and is defined as follows:

$$S = (U, C, D) \tag{5}$$

where C is a disjoint set of condition attributes and D is the decision attribute. For every $x \subset U$, there exist C1(x), ... , Cn(x), d1(x), ... , dm(x). The decision rule induced by x in S is:

$$C1(x), \ldots , Cn(x) \rightarrow d1(x), \ldots , dm(x) \text{ or } C \rightarrow D. \tag{6}$$

where the arrow implies the decision D is based on condition C. The importance degree of attributes relative to the decision is calculated as:

$$\gamma cd\ (c) = \{ | posc(D) | / | U | \} - \{ | pos(c - \{c\})(D) | / | U | \}. \tag{7}$$

This way the most important attributes are selected and if the importance degree equals zero, the attribute is unimportant. The larger γcd (c), the higher the attribute degree of importance is. Please note that the importance degree is not a percentage and has no units. If $|x|$ is the number of element in a set (i.e., cardinality of x), then the support of decision is defined as:

$$suppx(C,D) = | A(x) | = | C(x) \cap D(x) | \tag{8}$$

and the strength of the decision is quantified as:

$$\sigma x(C,D) = suppx(C,D) / | U | . \tag{9}$$

In other words, the support of the decision corresponds to the number of times that a certain rule is observed within the dataset and the strength of the decision is the support of the decision divided by the total number of decision rules. So, if the support of a decision is high, it means that the number of times that the specific decision rule is repeated is high and consequently, this decision rule is strong.

Also, the certainty of the decision rule is calculated as:

$$cerx(C,D) = [| C(x) \cap D(x) |] / | C(x) | = suppx(C,D) / | C(x) | = \sigma x(C,D) / \pi | C(x) | \tag{10}$$

where $\pi | C(x) | = | C(x) | / | U |$. When cerx equals to one, then $C \rightarrow xD$ is a certain decision rule.

Another useful factor in the DM rule concept is the coverage of decision rule defined as:

$$covx(C,D) = [| C(x) \cap D(x) |] / | D(x) | = suppx(C,D) / | D(x) | = \sigma x(C,D) / \pi | D(x) | \tag{11}$$

where $\pi | C(x) | = | D(x) | / | U |$. The coverage coefficient is the conditional probability of reasons for a given decision.

When $C \rightarrow xD$ is a decision rule, then $D \rightarrow xC$ is called the inverse decision rule and can be used to give explanations (reasons) for a decision. Please note that the certainty factors for inverse decision rules are coverage factors for the original decision rule [14].

2.2. Study Area and Dataset

In this study, we evaluate the chemical and physical quality of water at the outlet of the watershed that contains the George Mason University campus in Fairfax, VA. Figure 1 shows the watershed boundaries and the location where water quality indicators were sampled. This urbanized watershed contains two small creeks and one retention pond and is located within the larger Pohick Creek Watershed. Moreover, it consists of generally flat to sloping topography with most drainage (approximately 90%) flowing towards the south central portion of the campus and Pohick Creek.

A water quality monitoring instrument (the Eureka Manta2 Waterprobe) with six sensors automatically records six water quality indicators (listed in Table 1): dissolved oxygen concentration (DO), nitrate concentration, pH, specific conductivity, temperature, and turbidity. These water quality indicators were chosen for several reasons. First off, they are listed by the Environmental Protection Agency (EPA) to define water quality standards for surface water [27]. Secondly, since George

Mason University complies with the Clean Water Act and EPA storm water regulations, its Facilities Department monitors these indicators across the campus every year [28]. The water quality probe recorded each indicator every hour from October 2015 to December 2017. However, the probe was out for calibration and repairs occasionally and there have been some frequent network issues with the data logger. As a result, only 14 months of data are used in this work.

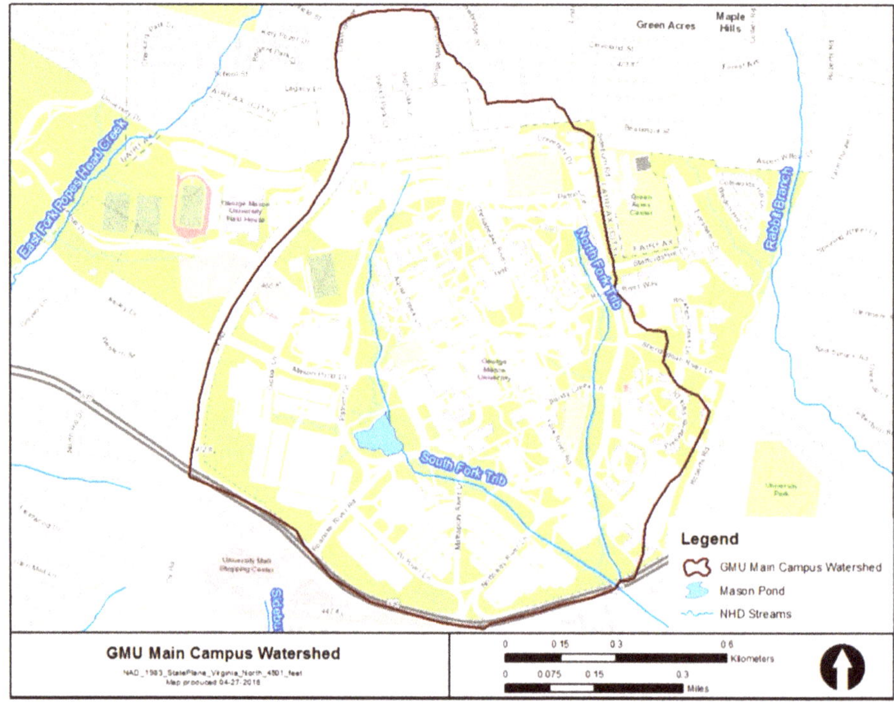

Figure 1. Location of study site.

Most data have been collected during 2016 and 2017. Both years are characterized by monthly mean temperature and precipitation that are similar to the 30-year mean values for the region [29]. Specifically, the average temperature during the past 30-year in Fairfax, VA is 13 °C and the yearly average temperature both in 2016 and 2017 is 14 °C. The average 30-year cumulative precipitation is 107 cm and the average precipitation for 2016 and 2017 is 90 cm and 104 cm, respectively. This indicates that 2016 and 2017 are not anomalous years in terms of regional climatology [29]. The collected data, summarized in Table 1, show that water temperature fluctuates from about 5 °C in winter to almost 30 °C in summer. The average pH is 6.75 and it falls in the range identified by EPA water quality standards for the Commonwealth of Virginia [27]. The average DO is 6.14 mg/L and it is also within the EPA water quality standards. The level of nitrate (average of 136.11 mg/L-N) shows that the runoff water possibly traveled through lands with fertilizers. Another possible source of nitrate is the atmosphere containing nitrogen compounds derived from automobiles [30]. According to EPA, the natural level of nitrate from wastewater effluent can range up to 30 mg/L. Finally, the high standard deviations in conductivity and turbidity are also common because of the frequent storms in this region.

The collected data are then discretized into three categories (low, medium and high): (i) any value lower than the 25th quartile is classified as low (L); (ii) any value between the 25th and 75th quartiles is classified as medium (M); and (iii) any value higher than the 75th quartile is classified as high (H).

Table 1. Units, average, standard deviation, 25th and 75th quartiles of water quality indicators collected during the study period at the location shown in Figure 1.

	Dissolved Oxygen	Nitrate Concentration	Specific Conductivity	Temperature	Turbidity	pH
Symbol	DO	NO$_3$	K	T	Tu	pH
Units	mg/L	mg/L	uS/cm	°C	NTU	-
Average	6.14	136	342	20.7	40.9	6.75
St. Deviation	1.23	46.4	126	3.63	95.7	0.16
25th Quartile	5.58	128	280	18.4	5.67	6.65
75th Quartile	6.72	158	384	23.3	36	6.82

A plot of time series of all the water quality indicators during the study period is shown in Figure 2. The inverse correlation between pH and temperature is clearly notable. However, it is important to mention that correlations between water quality indicators (as shown in [31]) at monthly scales are affected by several parameters, including environmental conditions and anthropogenic factors (e.g., rainfall events, construction sites).

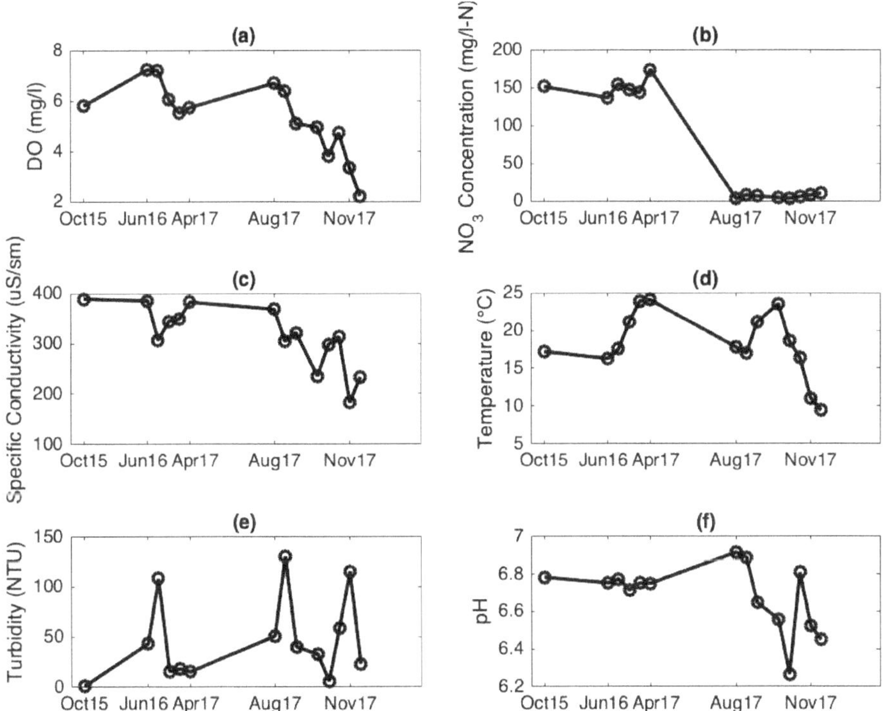

Figure 2. Time series for the six water quality indicators during the study period: (**a**) DO; (**b**) NO$_3$; (**c**) specific conductivity; (**d**) temperature; (**e**) turbidity; and (**f**) pH.

The shape of the frequency distributions for the water quality indicators considered in this study demonstrates the difficulty of fitting a known distribution to these datasets (Figure 3). For instance, the turbidity frequency distribution—shown in Figure 3e—is clearly non-normal and skewed towards lower values, with a long tail at higher values. On the other hand, some indicators (e.g., DO, specific conductivity) show more symmetrical distributions.

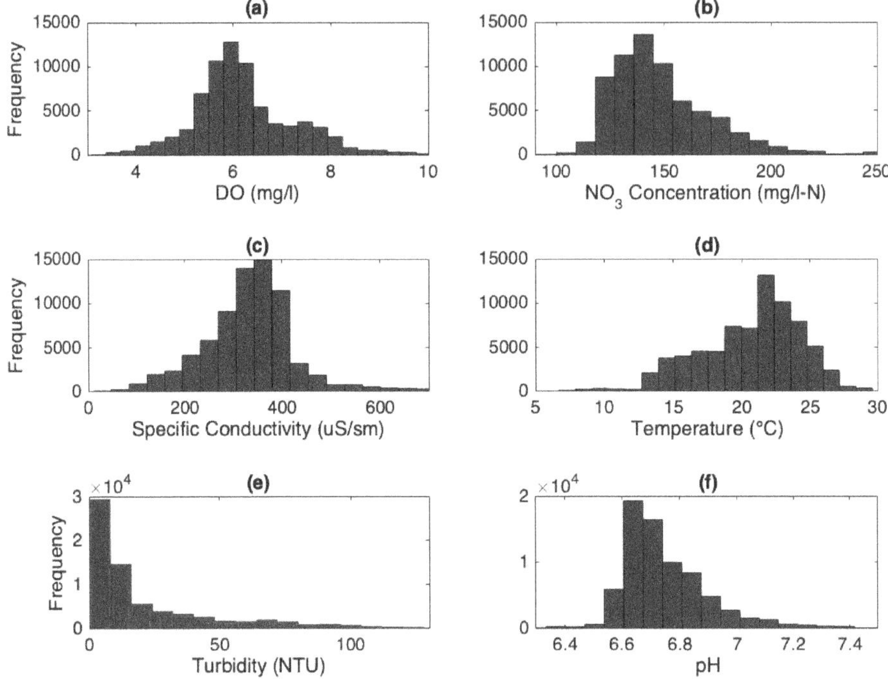

Figure 3. Frequency distribution histograms for: (**a**) DO; (**b**) NO$_3$; (**c**) specific conductivity; (**d**) temperature; (**e**) turbidity; and (**f**) pH during the study period.

3. Results

The information table is set up to apply RST to the data collected at the Mason watershed outlet. Specifically, five water quality indicators are chosen as the conditional attributes and the sixth one as the decision attribute. The water quality probe reads each water quality indicator every hour. However, in order to introduce rough set theory to water quality analysis, coarse resolution (monthly average) data are examined. This not only helps with showing a limited amount of condition and decision attributes in the following tables, but also helps to reduce the random noise in the data sample. Numerical values are assigned to each of the 14 months and presented as time codes in Table 2. The following analysis is based on the scenario in which pH is chosen as the decision attribute (D) and the rest of the indicators as condition attributes (C). In set theory formalism, this corresponds to U = {1, 2, 3, 4, 5, 6, 7, 8, 9, 10, 11, 12, 13, 14}, C = {DO, NO$_3$, K, T, Tu}, and D = {pH}, where C and D can be either L, or M, or H.

The first step is to identify redundant (or identical) time codes. After analyzing each time code, {7} and {8} are the only ones found identical, not only in terms of attributes, but also in terms of decision. This means that every single conditional and decision attribute is the same for time codes {7} and {8}. The fact that they are identical not only in terms of condition attributes but also in terms of decision attributes shows that if DO and K are medium, NO$_3$ and T are low, and Tu is high, pH is certainly high. This is the first certain decision rule concluded from Table 2. No other time codes are found to be identical in terms of condition and/or decision attributes. Thus, since all the other codes are unique in terms of both condition and decision attributes, each one of them represents a unique rule. As a result, 13 unique rules are identified in Table 2.

Table 2. Attributes and decision values where pH is the decision attributes and the other indicators are condition attributes.

Time Code	Date (M-Y)	DO	NO_3	K	T	Tu	pH
1	October-15	M	M	H	L	L	M
2	April-16	H	M	H	L	H	M
3	May-16	H	M	M	L	H	M
4	June-16	M	M	M	M	M	M
5	July-16	L	M	M	H	M	M
6	August-16	M	H	M	H	M	M
7	April-17	M	L	M	L	H	H
8	May-17	M	L	M	L	H	H
9	June-17	L	L	M	M	H	L
10	August-17	L	L	L	H	M	L
11	September-17	L	L	M	M	M	L
12	October-17	L	L	M	L	H	M
13	November-17	L	L	L	L	H	L
14	December-17	L	L	L	L	M	L

The second step explores the discernibility relation by eliminating one condition attribute at the time. There are 6 tables in Table 3 and each table except the first one is missing one attribute. Firstly, as discussed above, time codes {7} and {8} are identical and they are highlighted. Secondly, if DO, NO_3, and T were removed, discernibility would be the same, as shown in Table 3(b),(c),(e). As a result, these three attributes are deemed dispensable. Thirdly, if K and Tu were removed, new decision rules would appear. These new rules are highlighted in Table 3(d),(f) as well. In Table 3(d), time codes {2} and {3} are identical both in terms of condition and decision attributes, however, time codes {12} and {13} are just identical in terms of condition attribute. In Table 3(f), time codes {9} and {11} and time codes {13} and {14} are alike in terms of condition and decision attributes. Hence, there is a change in discernibility making both K and Tu indispensable.

The formal process of identifying dispensable attributes is further investigated in Table 4. The first column represents the attribute that is removed, whilst the second column represents the unique condition attribute combination in the absence of the corresponding attribute. When K and Tu are removed, the unique condition attribute combinations are different than when other indicators are removed in the other cases. In the third column, the unique decision making rules are displayed. If column 3 is not identical to the rules found in the presence of all attributes (conditional and decision), then the removed attribute is deemed dispensable (column 5). More specifically, the posc(D) is equal to {(1), (2), (3), (4), (5), (6), (7,8), (9), (10), (11), (12), (13), (14)}. If column 3 does not match posc(D), the removed attribute is indispensable. As a result of the analyses shown in Tables 3 and 4, the indispensable attributes are specific conductivity and turbidity. Clearly, in the absence of K, decision rules {2} and {3} are identical, in the absence of Tu, decision rules {9} and {11} are identical, and decision rules {13} and {14} are also identical. These attributes are defined as the core attributes.

Table 3. Analysis of the discernibility relation with identical time code highlighted for the following cases: (a) all attributes; (b) DO eliminated; (c) NO_3 concentration eliminated; (d) specific conductivity eliminated; (e) temperature eliminated; (f) turbidity eliminated.

(a)

Time Code	DO	NO_3	K	T	Tu	pH
1	M	M	H	L	L	M
2	H	M	H	L	H	M
3	H	M	M	L	H	M
4	M	M	M	M	M	M
5	L	M	M	H	M	M
6	M	H	M	H	M	M
7	M	L	M	L	H	H
8	M	L	M	L	H	H
9	L	L	M	M	H	L
10	L	L	L	H	M	L
11	L	L	M	M	M	L
12	L	L	M	L	H	M
13	L	L	L	L	H	L
14	L	L	L	L	M	L

(b)

Time Code	DO	NO_3	K	T	Tu	pH
1		M	H	L	L	M
2		M	H	L	H	M
3		M	M	L	H	M
4		M	M	M	M	M
5		M	M	H	M	M
6		H	M	H	M	M
7		L	M	L	H	H
8		L	M	L	H	H
9		L	M	M	H	L
10		L	L	H	M	L
11		L	M	M	M	L
12		L	M	L	H	M
13		L	L	L	H	L
14		L	L	L	M	L

(c)

Time Code	DO	NO_3	K	T	Tu	pH
1	M		H	L	L	M
2	H		H	L	H	M
3	H		M	L	H	M
4	M		M	M	M	M
5	L		M	H	M	M
6	M		M	H	M	M
7	M		M	L	H	H
8	M		M	L	H	H
9	L		M	M	H	L
10	L		L	H	M	L
11	L		M	M	M	L
12	L		M	L	H	M
13	L		L	L	H	L
14	L		L	L	M	L

(d)

Time Code	DO	NO_3	K	T	Tu	pH
1	M	M		L	L	M
2	H	M		L	H	M
3	H	M		L	H	M
4	M	M		M	M	M
5	L	M		H	M	M
6	M	H		H	M	M
7	M	L		L	H	H
8	M	L		L	H	H
9	L	L		M	H	L
10	L	L		H	M	L
11	L	L		M	M	L
12	L	L		L	H	M
13	L	L		L	H	L
14	L	L		L	M	L

(e)

Time Code	DO	NO_3	K	T	Tu	pH
1	M	M	H		L	M
2	H	M	H		H	M
3	H	M	M		H	M
4	M	M	M		M	M
5	L	M	M		M	M
6	M	H	M		M	M
7	M	L	M		H	H
8	M	L	M		H	H
9	L	L	M		H	L
10	L	L	L		M	L
11	L	L	M		M	L
12	L	L	M		H	M
13	L	L	L		H	L
14	L	L	L		M	L

(f)

Time Code	DO	NO_3	K	T	Tu	pH
1	M	M	H	L		M
2	H	M	H	L		M
3	H	M	M	L		M
4	M	M	M	M		M
5	L	M	M	H		M
6	M	H	M	H		M
7	M	L	M	L		H
8	M	L	M	L		H
9	L	L	M	M		L
10	L	L	L	H		L
11	L	L	M	M		L
12	L	L	M	L		M
13	L	L	L	L		L
14	L	L	L	L		L

Table 4. Calculating the discernibility and dispensability.

Attribute C	U/Ind(C-{c})	Pos(c-{c})(D)	Pos(c-{c})(D) = Posc(D)?	Indispensability
DO	(1), (2), (3), (4), (5), (6), (7,8), (9), (10), (11), (12), (13), (14)	(1), (2), (3), (4), (5), (6), (7,8), (9), (10), (11), (12), (13), (14)	Y	N
NO$_3$	(1), (2), (3), (4), (5), (6), (7,8), (9), (10), (11), (12), (13), (14)	(1), (2), (3), (4), (5), (6), (7,8), (9), (10), (11), (12), (13), (14)	Y	N
K	(1), (2,3), (4), (5), (6), (7,8), (9), (10), (11), (12,13), (14)	(1), (2,3), (4), (5), (6), (7,8), (9), (10), (11), (12), (13), (14)	N	Y
T	(1), (2), (3), (4), (5), (6), (7,8), (9), (10), (11), (12), (13), (14)	(1), (2), (3), (4), (5), (6), (7,8), (9), (10), (11), (12), (13), (14)	Y	N
Tu	(1), (2), (3), (4), (5), (6), (7,8), (9,11), (10), (12), (13,14)	(1), (2), (3), (4), (5), (6), (7,8), (9,11), (10), (12), (13,14)	N	Y

The application of RST to the water quality data sampled at the Mason campus that was discussed above identified the fundamental and redundant water quality indicators. Their importance degree as conditional attributes in determining the other indicators (decision) is then quantified using Equation (7). Table 5 shows the indispensable attributes for each decision attribute and their corresponding importance degree. There are only two indispensable attributes identified for each decision attribute. Specifically, two indispensable attributes are identified for pH, DO, T, and K, whereas only one indispensable attribute is identified for Tu (i.e., specific conductivity). This also means that several attributes are redundant and not necessary to fill in possible gaps in time series. This kind of conclusion is extremely useful when obtaining ground observations is complicated by impervious terrain, financing constraints, and/or extreme atmospheric conditions.

Results in Table 5 demonstrate that turbidity is equally important in every scenario considered in the study with an importance degree of 0.14. Specific conductivity is the next important factor with an importance degree of 0.07. If the decision attribute is turbidity, the only indispensable attribute is specific conductivity and if the specific conductivity is the decision attribute, the indispensable attribute beside turbidity is temperature. According to the foregoing analysis, if any of the six water quality indicator needs to be retrieved because of a missed measurement, turbidity and specific conductivity are the core values that would provide useful information about the missing information. Moreover, the decision in every scenario is weighted towards turbidity since the importance degree of turbidity is higher than the importance degree of conductivity.

Table 5. Importance degree of C attributes relative to the decision attribute D.

Decision Attribute	Indispensable Attribute 1 (Importance Degree)	Indispensable Attribute 2 (Importance Degree)
pH	Tu (0.14)	K (0.07)
DO	Tu (0.14)	K (0.07)
NO$_3$	Tu (0.14)	K (0.07)
T	Tu (0.14)	K (0.07)
Tu	K (0.07)	-
K	Tu (0.14)	T (0.07)

There is a direct relationship between temperature and all the other water quality indicators. Furthermore, conductivity has an effect on turbidity and turbidity influences dissolved oxygen concentration, which also affects nitrate concentration. However, there is no direct relation between pH and other indicators. Therefore, we start our analysis by selecting pH as a decision attribute. Based on the dispensability analyses shown above, conductivity and turbidity are the core condition attributes (Table 6). There is a strong relationship between these two core attributes in stormwater runoff across the Mason campus watershed, as previously shown by [32].

Table 6. Indicators of decision-making (DM) rules.

Decision Rule	K	Tu	pH	N	Strength	Certainty	Coverage
1	H	L	M	1	0.07	1	0.14
2	H	H	M	1	0.07	1	0.14
3	M	H	M	2	0.14	0.4	0.29
4	M	M	M	3	0.21	0.75	0.43
5	M	H	H	2	0.14	0.4	1.00
6	M	H	L	1	0.07	0.2	0.20
7	L	M	L	2	0.14	1	0.40
8	M	M	L	1	0.07	0.25	0.20
9	L	H	L	1	0.07	1	0.20

Table 6 shows the DM rules together with their strength, certainty, and coverage, computed according to Equations (9)–(11), respectively. Table 6 also shows the support of each DM rule (N). As mentioned above, N is the number of times that each DM rule was recorded. Table 6 shows that N is larger than 1 for DM rules 3, 4, 5, and 7. As a result, their strengths are higher than the strengths of the rules for which N = 1.

If the conditional attributes are identical and the decision attributes are not equal, the certainty of the DM rule is less than one. Thus, the certain DM rules are 1, 2, 7, and 9. In other words, if specific conductivity is high and turbidity is either low or high, then pH is certainly medium (according to DM rule 1 and 2). If specific conductivity is low and turbidity is either medium or high, then pH is certainly low (according to DM rule 7 and 9).

In order to explain the decision attribute in terms of condition attributes, the conditions and decision attributes need to be mutually replaced in every DM rule. The only certain inverse rule is DM rule 5, which indicates that if pH is high, then turbidity is high and specific conductivity is medium. Moreover, rule number 5 is a unique case. Since there is only one rule with a high pH value, the coverage for this rule is equal to 1 and, as a result, the certainty for inverse DM rule 5 is one.

The same analysis is repeated five times by selecting a different attribute as a decision attribute and setting the rest of the attributes as a condition attributes every time. Table A1 shows the DM rules and strength, certainty, and coverage for all the other cases. The highest strength factor (0.29) belongs to the rule in which the conditional attributes are specific conductivity and turbidity and the decision attribute is temperature. On the other hand, five rules show a certainty factor equal to 1 when the conditional attributes are specific conductivity and turbidity and the decision attribute is dissolved oxygen. Moreover, the coverage factor equals to 1 in one of the rules when the specific conductivity and turbidity are conditional attributes and the nitrate is decision attribute.

A similar analysis was performed also at weekly scale, by averaging the water quality indicators for each week of the study period. However, because of the high temporal variability in water quality, no redundant attribute was identified. Hence, at finer temporal resolutions, more attributes play an important role. Since this work is meant as an attempt to apply rough set theory to water quality data analysis, it would not be feasible to effectively display the step-by-step procedure using a larger dataset (e.g., weekly). Nevertheless, the developed approach based on rough set theory could be applied to data at any temporal resolution and to time series of any length.

The developed methodology can also be used to compare different months or the same month in different years. For instance, the months of April, May, June, and August of 2016 (case 1) can be compared to the same months in 2017 (case 2). In case 1, the indispensable attribute would be specific conductivity. However, in case 2, there is no indiscernible attribute. This shows that indiscernible attributes may vary depending on environmental and/or anthropogenic conditions. This kind of comparison highlights possible changes in the stream water quality conditions, whose sources can be potentially investigated by the analyst.

4. Discussion

This study investigates the application of RST to water quality analysis. RST does not require any prior information on the dataset and represents a powerful tool to deal with uncertainty and vagueness in the sample. Moreover, RST is capable of finding indiscernible attributes and extracting rules based on core attributes. This work presents the basic concepts of rough set theory and its application to six water quality indicators collected during a 3-year-long study period at the George Mason University campus in Fairfax, VA. More specifically, monthly averages for each water quality indicator are calculated and 14 months are considered.

It is important to mention that the streamflow velocity at the watershed outlet where data were collected is particularly high during and after rainfall events. As a result, the common relationships among water quality attributes are not observed in this case study that focuses on the monthly scale. For example, when water temperature is low, DO concentration is commonly high [33]. However, we cannot observe this rule at the monthly resolution. When a storm happens, even during summer when temperatures are high, the rapidly moving water contains more DO than stagnant water in winter days (when the temperature is lower).

Coarse temporal resolution (i.e., monthly) data are selected here in order to present a novel methodology in the field of water quality analysis. The coarse resolution helps with showing a limited number of attributes and decision values. Six different scenarios are studied here and in each scenario one attribute is assigned to be a decision attribute and the rest are reflected as conditional attributes. In most cases, specific conductivity (with an importance degree of 0.07) and turbidity (with an importance degree of 0.14) are the core conditional attributes. In addition, we generate DM rules for each scenario and calculate the strength, certainty, and coverage of each rule. The certain rules show that if specific conductivity is high and turbidity is either low or high, then pH is medium. Also, if specific conductivity is low and turbidity is either medium or high, then pH is certainly low. However, the coverage of these DM rules is the lowest among all DM rules. Five other possible DM rules with certainty lower than one are identified as well. There is one DM rule with coverage factor of one (DM 5), which means that there is only one DM rule with a unique pH value (high). As a result, the certainty for the inverse DM rule 5 is one.

Overall, RST was proven capable of finding core indicators and discovering DM rules. Considering more attributes and more data entries could increase the certainty of the identified DM rules and possibly identify additional DM rules. RST-based DM rules can be of tremendous help to planners and analysts in their decision making process. For instance, results from this study can be useful for university facility managers that monitor water quality across campus. If applied to a larger scale, the proposed methodology has the potential of providing timely, relevant, and essential water quality information.

Future work should look at the raw data at their native resolution (one hour). Although no difference in the DM rules was observed in the weekly analysis with respect to the monthly one, increasing the resolution to one hour may result in higher certainty in the DM rules. Moreover, other locations should be investigated to verify the efficiency of the proposed methodology and possibly sampling additional indicators (i.e., conditional attributes). Further conditional attributes can be related to atmospheric conditions, like the amount and duration of precipitation events and land cover/land use.

Author Contributions: Conceptualization, M.Z. and V.M.; Data, M.Z.; Formal analysis, M.Z.; Methodology, M.Z. and V.M.; Project administration, M.Z. and V.M.; Resources, V.M.; Supervision, V.M.; Writing—original draft, M.Z.; Writing—review & editing, V.M.

Funding: This research received no external funding.

Conflicts of Interest: The authors declare no conflict of interest.

Appendix A

Table A1. DM rules for various scenarios, with different decision attributes: (a) DO; (b) Tu; (c) NO_3; (d) T; and (e) k.

(a)

K	Tu	DO	N	Strength	Certainty	Coverage
H	L	M	1	0.07	1.00	0.20
H	H	H	1	0.07	1.00	0.50
M	H	H	1	0.07	1.00	0.50
M	M	M	2	0.14	0.50	0.40
M	M	L	2	0.14	0.50	0.29
M	H	M	2	0.14	0.40	0.40
M	H	L	2	0.14	0.40	0.40
L	M	L	2	0.14	1.00	0.40
L	H	L	1	0.07	1.00	0.20

(b)

DO	K	Tu	N	Strength	Certainty	Coverage
H	H	H	1	0.07	1.00	0.14
H	M	H	1	0.07	1.00	0.14
L	M	H	2	0.14	0.50	0.29
L	L	H	1	0.07	0.33	0.14
M	M	H	2	0.14	0.50	0.29
M	H	L	1	0.07	1.00	1.00
L	M	M	2	0.14	0.50	0.33
L	L	M	2	0.14	0.67	0.33
M	M	M	2	0.14	0.50	0.33

(c)

K	Tu	NO_3	N	Strength	Certainty	Coverage
M	M	H	1	0.07	0.25	1.00
M	H	L	2	0.14	0.40	0.25
L	M	L	2	0.14	1.00	0.25
M	H	L	1	0.07	0.20	0.13
L	H	L	1	0.07	1.00	0.13
M	H	L	1	0.07	0.20	0.13
M	M	L	1	0.07	0.25	0.13
H	L	M	1	0.07	1.00	0.20
H	H	M	1	0.07	1.00	0.20
M	H	M	1	0.07	0.20	0.20
M	M	M	2	0.14	0.50	0.40

(d)

K	Tu	T	N	Strength	Certainty	Coverage
L	M	H	1	0.07	0.50	0.33
M	M	H	2	0.14	0.50	0.67
H	L	L	1	0.07	1.00	0.13
H	H	L	1	0.07	1.00	0.13
L	H	L	1	0.07	1.00	0.13
L	M	L	1	0.07	0.50	0.13
M	H	L	4	0.29	0.80	0.50
M	M	M	2	0.14	0.50	0.67
M	H	M	1	0.07	0.20	0.33

(e)

T	Tu	K	N	Strength	Certainty	Coverage
L	L	H	1	0.07	1.00	0.50
L	H	H	1	0.07	0.17	0.50
H	M	L	1	0.07	0.33	0.33
L	H	L	1	0.07	0.17	0.33
L	M	L	1	0.07	1.00	0.33
M	M	M	2	0.14	1.00	0.22
H	M	M	2	0.14	0.67	0.22
L	H	M	3	0.21	0.50	0.33
M	H	M	1	0.07	1.00	0.11
L	H	M	1	0.07	0.17	0.11

References

1. Pai, P.-F.; Li, L.-L.; Hung, W.-Z.; Lin, K.-P. Using ADABOOST and Rough Set Theory for Predicting Debris Flow Disaster. *Water Resour. Manag.* **2014**, *28*, 1143–1155. [CrossRef]
2. Wagner, R.J.; Boulger, R.W., Jr.; Oblinger, C.J.; Smith, B.A. Guidelines and Standard Procedures for Continuous Water-Quality Monitors: Station Operation, Record Computation, and Data Reporting [Internet]. 2006 [Cited 1 June 2018]. (Techniques and Methods). Report No.: 1-D3. Available online: http://pubs.er.usgs.gov/publication/tm1D3 (accessed on 5 November 2018).
3. Lepot, M.; Aubin, J.-B.; Clemens, F.H.L.R. Interpolation in Time Series: An Introductive Overview of Existing Methods, Their Performance Criteria and Uncertainty Assessment. *Water* **2017**, *9*, 796. [CrossRef]
4. Fu, L.; Wang, Y.-G. Statistical Tools for Analyzing Water Quality Data | IntechOpen [Internet]. 2012 [Cited 3 July 2018]. Available online: /books/water-quality-monitoring-and-assessment/statistical-tools-for-analyzing-water-quality-data (accessed on 5 November 2018).
5. Liou, S.M.; Lo, S.L.; Hu, C.Y. Application of two-stage fuzzy set theory to river quality evaluation in Taiwan. *Water Res.* **2003**, *37*, 1406–1416. [CrossRef]
6. Chen, X.; Li, Y.S.; Liu, Z.; Yin, K.; Li, Z.; Wai, O.W.; King, B. Integration of multi-source data for water quality classification in the Pearl River estuary and its adjacent coastal waters of Hong Kong. *Cont. Shelf Res.* **2004**, *24*, 1827–1843. [CrossRef]

7. Shrestha, S.; Kazama, F. Assessment of surface water quality using multivariate statistical techniques: A case study of the Fuji river basin, Japan. *Environ. Model. Softw.* **2007**, *22*, 464–475. [CrossRef]

8. Diamantopoulou, M.J.; Antonopoulos, V.Z.; Papamichail, D.M. Cascade Correlation Artificial Neural Networks for Estimating Missing Monthly Values of Water Quality Parameters in Rivers. *Water Resour. Manag.* **2007**, *21*, 649–662. [CrossRef]

9. Singh, A.P.; Ghosh, S.K.; Sharma, P. Water quality management of a stretch of river Yamuna: An interactive fuzzy multi-objective approach. *Water Resour. Manag.* **2007**, *21*, 515–532. [CrossRef]

10. Manache, G.; Melching, C.S. Identification of reliable regression- and correlation-based sensitivity measures for importance ranking of water-quality model parameters. *Environ. Model. Softw.* **2008**, *23*, 549–562. [CrossRef]

11. Qin, X.S.; Huang, G.H. An Inexact Chance-constrained Quadratic Programming Model for Stream Water Quality Management. *Water Resour Manag.* **2009**, *23*, 661. [CrossRef]

12. Hou, D.; He, H.; Huang, P.; Zhang, G.; Loaiciga, H. Detection of water-quality contamination events based on multi-sensor fusion using an extended Dempster–Shafer method. *Meas. Sci. Technol.* **2013**, *24*, 055801. [CrossRef]

13. Pawlak, Z. Rough sets. *Int. J. Comput. Inf. Sci.* **1982**, *11*, 341–356. [CrossRef]

14. Polkowski, L. Rough Sets: Mathematical Foundations [Internet]. Physica-Verlag Heidelberg. 2002 [Cited 27 April 2018]. (Advances in Intelligent and Soft Computing). Available online: //www.springer.com/us/book/9783790815108 (accessed on 5 November 2018).

15. Skowron, A.; Suraj, Z. *Rough Sets and Intelligent Systems—Professor Zdzisław Pawlak in Memoriam*; Springer Science & Business Media: Berlin, Germany, 2012; p. 682.

16. Nguyen, T.-T.; Nguyen, P.-K. Reducing Attributes in Rough Set Theory with the Viewpoint of Mining Frequent Patterns. *Int. J. Adv. Comput. Sci. Appl.* **2013**, *4*. [CrossRef]

17. Pawlak, Z. Rough set theory and its applications. *J. Telecommun. Technol.* **2002**, *3*, 7–10.

18. Dong, S.-H.; Zhou, H.-C.; Xu, H.-J. A Forecast Model of Hydrologic Single Element Medium and Long-Period Based on Rough Set Theory. *Water Resour. Manag.* **2004**, *18*, 483–495. [CrossRef]

19. Pai, P.-F.; Lee, F.-C. A Rough Set Based Model in Water Quality Analysis. *Water Resour. Manag.* **2010**, *24*, 2405–2418. [CrossRef]

20. Shen, Q.; Chouchoulas, A. FuREAP: A Fuzzy–Rough Estimator of Algae Populations. *Artif. Intell. Eng.* **2001**, *15*, 13–24. [CrossRef]

21. Barbagallo, S.; Consoli, S.; Pappalardo, N.; Greco, S.; Zimbone, S.M. Discovering Reservoir Operating Rules by a Rough Set Approach. *Water Resour. Manag.* **2006**, *20*, 19–36. [CrossRef]

22. Predki, B.; Słowiński, R.; Stefanowski, J.; Susmaga, R.; Wilk, S. ROSE—Software Implementation of the Rough Set Theory. In *Rough Sets and Current Trends in Computing [Internet]*; Lecture Notes in Computer Science; Springer: Berlin/Heidelberg, Germany, 1998; pp. 605–608.

23. Ip, W.C.; Hu, B.Q.; Wong, H.; Xia, J. Applications of rough set theory to river environment quality evaluation in China. *Water Resour.* **2007**, *34*, 459–570. [CrossRef]

24. Karami, J.; Alimohammadi, A.; Seifouri, T. Water quality analysis using a variable consistency dominance-based rough set approach. *Comput. Environ. Urban Syst.* **2014**, *43*, 25–33. [CrossRef]

25. Pawlak, Z.; Grzymala-Busse, J.; Slowinski, R.; Ziarko, W. Rough Sets. *Commun. ACM* **1995**, *38*, 88–95. [CrossRef]

26. Rissino, S.; Lambert-Torres, G. Rough Set Theory—Fundamental Concepts, Principals, Data Extraction, and Applications. In *Data Mining and Knowledge Discovery in Real Life Applications*; Ponce, J., Karahoca, A., Eds.; IN-TECH: Hong Kong, China, 2009; pp. 35–58.

27. Department, V.; Quality, E. Virginia Administrative Code, Title 9. Environment, Agency 25. State Water Control Board, Chapter 260. Water Quality Standards. Available online: https://www.epa.gov/sites/production/files/2014-12/documents/vawqs.pdf (accessed on 5 November 2018).

28. Mason MS4 Program | Facilities | George Mason University [Internet]. [Cited 24 October 2018]. Available online: https://facilities.gmu.edu/resources/land-development/ms4/ (accessed on 5 November 2018).

29. NWSCIW. National Weather Service Sterling [Internet]. [Cited 27 October 2018]. Available online: https://w2.weather.gov/climate/local_data.php?wfo=lwx (accessed on 5 November 2018).

30. Nitrogen and Water: USGS Water Science School [Internet]. [Cited 27 July 2018]. Available online: https://water.usgs.gov/edu/nitrogen.html (accessed on 5 November 2018).

31. Copetti, D.; Marziali, L.; Viviano, G.; Valsecchi, L.; Guzzella, L.; Capodaglio, A.G.; Tartari, G.; Polesello, S.; Valsecchi, S.; Mezzanotte, V.; et al. Intensive monitoring of conventional and surrogate quality parameters in a highly urbanized river affected by multiple combined sewer overflows. *Water Sci. Technol. Water Suppl.* **2018**. [CrossRef]
32. Gholoom, A. Studying the Impact of Different Green Rooftop Designs on Stormwater [Internet] [Thesis]. 2018 [Cited 17 July 2018]. Available online: http://mars.gmu.edu/handle/1920/10916 (accessed on 5 November 2018).
33. Dissolved Oxygen, from the USGS Water Science School: All about Water. [Internet]. [Cited 26 July 2018]. Available online: https://water.usgs.gov/edu/dissolvedoxygen.html (accessed on 5 November 2018).

Article

Use of the WRF-DA 3D-Var Data Assimilation System to Obtain Wind Speed Estimates in Regular Grids from Measurements at Wind Farms in Uruguay

Gabriel Cazes Boezio [1,*] **and Sofía Ortelli** [1,2]

[1] School of Engineeering, Universidad de la República, Montevideo 11300, Uruguay; sofiao@fing.edu.uy
[2] Usinas y Trasmisiones del Estado (UTE), Montevideo 11200, Uruguay
* Correspondence: agcm@fing.edu.uy; Tel.: +598-27115279

Received: 31 August 2019; Accepted: 24 October 2019; Published: 29 October 2019

Abstract: This work assessed the quality of wind speed estimates in Uruguay. These estimates were obtained using the Weather Research and Forecast Model Data Assimilation System (WRF-DA) to assimilate wind speed measurements from 100 m above the ground at two wind farms. The quality of the estimates was assessed with an anemometric station placed between the wind farms. The wind speed estimates showed low systematic errors at heights of 87 and 36 m above the ground. At both levels, the standard deviation of the total errors was approximately 25% of the mean observed speed. These results suggested that the estimates obtained could be of sufficient quality to be useful in various applications. The assimilation process proved to be effective, spreading the observational gain obtained at the wind farms to lower elevations than those at which the assimilated measurements were taken. The smooth topography of Uruguay might have contributed to the relatively good quality of the obtained wind estimates, although the data of only two stations were assimilated, and the resolution of the regional atmospheric simulations employed was relatively low.

Keywords: data assimilation; 3D-Var

---

1. Summary

This work evaluated the use of techniques for assimilation of data from field measurements into initial conditions of atmospheric numerical simulations in order to obtain wind estimates in Uruguay, at heights of 100 m above the ground and lower. The wind was estimated with hourly frequency in a regular grid that covers the country. The field data to be assimilated was operatively measured in wind farms installed in Uruguay, using anemometers placed 100 m above the ground. The data was assimilated into initial conditions for the Weather Research and Forecast regional model (WRF) of the National Center of Atmospheric Research (NCAR), [1] using the module for data assimilation included in this model, the WRF-DA module [2].

The data assimilation process, also called analysis, is an essential component of numerical atmospheric forecasts, and its main purpose is to generate initial conditions for the predictions. The variables that compose an initial condition are called prognostic variables because the model uses their values at a given instant to compute their values at a later time. To generate an initial condition for a specific numerical model at a given time, a first approximation is generally used. This first approximation, called "background condition", usually consists of a prediction for the same instant, obtained with the same model, from previous initial conditions. The data assimilation system must combine the information from the background condition with the information from measurements of atmospheric variables (or variables of systems related to it; for example, the ocean, the soil, or the cryosphere). This combination of information is done in a way that optimizes the quality of the result in statistical terms, either minimizing the expected value of the sum of its quadratic errors or maximizing its likelihood. Note that the short-term predictions used as background values are, in turn, affected by field measurements that were assimilated during previous times. This allows considering information from regions or atmospheric levels in which few measurements are available since measurements, done earlier in other regions or at other levels, can propagate their influence through atmospheric dynamical processes into the zones that have relatively fewer observations.

Kalnay [3] described the main techniques for data assimilation that are currently used, such as optimal interpolation, 3D-Var, Kalman filters, ensemble Kalman filters, and 4D-Var. The WRF-DA system implements the 3D-Var [4] and 4D-Var techniques [5], and a technique that combines the ensemble Kalman filter with 3D-Var [6,7]. Harlim [8] pointed that the referred techniques assume that the errors from the background conditions and the field measurements are unbiased and normally distributed, although numerical models inevitably have systematic errors. These assumptions can provide reasonable estimates of the first-order statistics while being practical to implement. However, these methodologies require caution for interpreting their higher-order statistical estimates, and an important challenge in data assimilation is to use existing methods in the presence of model systematic errors. The author proposed methodologies to mitigate the effects of model systematic errors on the results of the assimilation process. Rao and Sandu [9] proposed an a-posteriori error estimation methodology that quantifies the impact of model and data errors on the inference results of inverse problems, including the 4D-Var assimilation process. The model and data errors considered include both unbiased noise and systematic biases. The authors found that the proposed methodology could be useful to reduce and quantify uncertainties in a real-time system with feedback. Besides, the error estimates can be used to locate faulty sensors and to determine areas of maximum sensitivity, where improvements in the stations network or an increase in model resolution may be required.

In addition to its direct use in the numerical prediction process, the results of data assimilation can be considered "pseudo-observations" of atmospheric variables in regular grids. Note that these do not consist purely of interpolations of field measurements in a regular grid since they also consider the information from short-term predictions. The uses of the pseudo-observations obtained from a data assimilation process can be very broad, but they require an evaluation of their quality by comparison with field measurements not used in the assimilation process.

The current work used the rather conventional 3D-Var assimilation technique. In Section 2, we described the data used, both from field measurements and from numerical predictions used as background conditions, and we referred to the quality control method used for the field data from the wind farms. In Section 3, we described the main aspects of the 3D-Var assimilation technique and its implementation in this work using the WRF DA system. In Section 4, we described the main results, and in Section 5, we presented the conclusions.

2. Data Description

The wind data used in the data assimilation process was obtained from two anemometers installed in the "Rosendo Mendoza" (WF1) and "Valentines" (WF2) wind farms. The geographic locations of these wind farms are shown in Figure 1. The anemometers were placed 100 m above the ground, and they recorded mean velocities for successive periods of 15 min, which were transmitted to the National Dispatch of Electric Charges of Uruguay, operated by the Electricity Market Administration (Administración del Mercado Eléctrico; ADME) and the National Administration of Electric Power Plants and Transmissions (Usinas y Trasmisiones del Estado; UTE), and Uruguayan National Public Electricity Utility Organizations. In addition to the wind measurements, the mean electric power generated in the same 15-min periods and the corresponding quantity of aerogenerators that were effectively active were also transmitted by each wind farm. These additional data allowed for a quality control analysis of the measurements, as described by Orteli and Cazes Boezio [10]. With the historical information of wind velocity and electric power generated, the authors built an empirical wind-power curve for each wind farm. In those cases in which not all the aerogenerators of the wind farm were active, the power generation that would correspond to a condition of full availability was estimated by linear extrapolation. The authors considered that those combinations of wind and generated power that depart from the empirical curve beyond certain thresholds were suspicious of being affected by malfunctioning of the measurement or recording systems, or by occasional interference of the wake of an aerogenerator with the anemometer. The data from WF1 and WF2 available for this study were from the months from January to May and from November to December of 2017 (seven months).

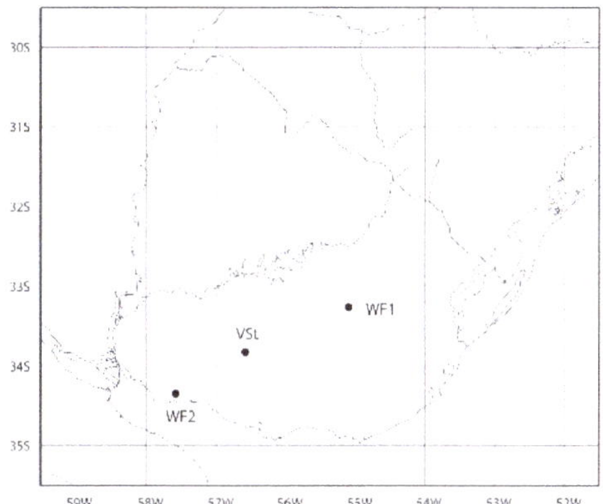

Figure 1. The geographic location of anemometers used for data assimilation (WF1 and WF2) and validation (Vst).

To evaluate the quality of the pseudo-observations obtained through the implemented data assimilation process, an independent station was used ("validation station", VSt), which is located at Colonia Arias, Uruguay. The station location is also shown in Figure 1. The station measures wind speed and direction at heights of 87 and 36 m above the ground and is part of a network of stations that measures wind velocity and solar radiation that has been operated by UTE since 2009. This network is described by Cornalino and Draper [11], and its measurements (that include wind velocities averaged every 10 min) are made available online by UTE. In recent years, many of the anemometric stations operated by UTE have become affected by wind farms. The VSt of Colonia Arias has been active since 2011, and during the period studied in this work, it was not affected by any wind farm. Figure 2 shows the wind speed at the WF1, WF2, and Vst sensors averaged for each local hour during the seven months considered.

The assimilation process uses regional simulations computed with the WRF model from NCAR, as shown by Cazes Boezio and Orteli [12]. The regional simulations take their initial and boundary conditions from global predictions made by the Global Forecast System (GFS) of the National Ocean and Atmosphere Administration (NOAA) of the Unites States. The horizontal grid of the regional simulations has a resolution of 30 km in the zonal and meridional directions, as shown in Figure 3. The vertical direction is discretized in 54 levels, 7 of which are within the first 100 m of height above the ground. In Appendix A, we indicated the parameterizations of physical processes employed, and we defined in detail the vertical discretization. The regional simulations have two purposes: first, they generate the background conditions into which the data measured in WF1 and WF2 are assimilated; and second, they allow for the estimation of the matrix of covariances of the errors of these background conditions. In Section 3, the hours of initialization and the simulated periods used for each one of these purposes were specified.

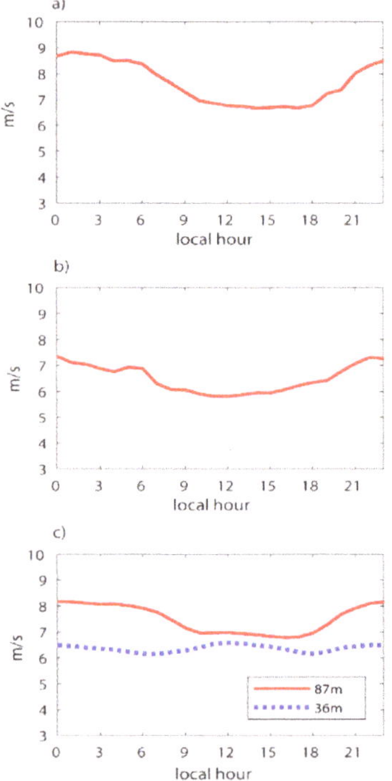

Figure 2. Mean wind velocity, as a function of the local hour, for the WF1 (**a**), WF2 (**b**), and VSt stations (**c**). At the VSt station, the solid line shows the mean velocity for the anemometer 87 m above the ground, and the dotted line shows the anemometer 36 m above the ground.

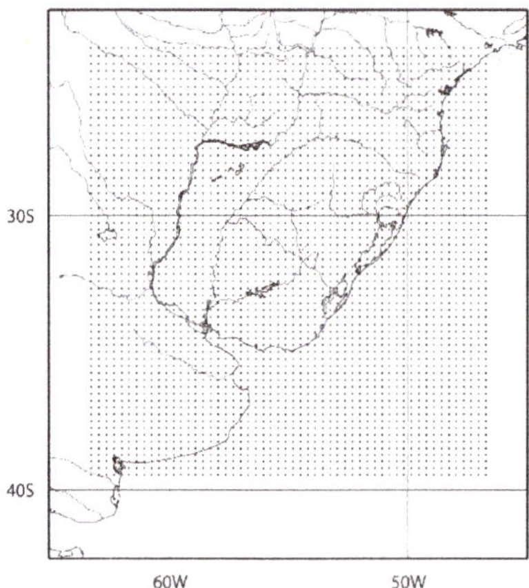

Figure 3. Grid points of the horizontal discretization used in the regional simulations.

3. Methods

Next, we described the main aspects of the 3D-Var assimilation technique, which was used in this work and is available in the WRF-DA system. A complete description of this technique and its implementation in the WRF-DA system can be found in the NCAR Technical Note 453 [13], in

Barker et al. [4] and Barker et al. [2], while the corresponding operational details are given in the WRF-DA User Guide [14].

The background condition information is displayed as a vector called $\mathbf{x_b}$, which contains the values of all the different prognostic variables that compose the background condition in a certain conventional order. This information is displayed for each variable and each point of the grid. The field measurements that are to be assimilated, obtained at the same instant that corresponds to $\mathbf{x_b}$, are displayed as a vector called $\mathbf{y}$, also in a conventional order. The initial condition determined by the assimilation system is expressed as a vector called $\mathbf{x_a}$, with a structure analogous to that of $\mathbf{x_b}$. $\mathbf{x_a}$ combines the information from the background condition $\mathbf{x_b}$ and the field measurements contained in $\mathbf{y}$, as expressed in Section 1. The 3D-Var technique determines the vector $\mathbf{x_a}$ as that of maximum likelihood, conditioned to the information of the background condition and the field measurements, and is the vector $\mathbf{x}$ that minimizes the following expression [3]:

$$J(x) = \frac{1}{2}(x - x_b)^T B^{-1}(x - x_b) + \frac{1}{2}(y - H(x))^T R^{-1}(y - H(x)) \tag{1}$$

that is, $J(x_a)$ must be minim. In this formula, $\mathbf{B}$ and $\mathbf{R}$ are the matrices of error covariances of the background condition $\mathbf{x_b}$ and the field measurements $\mathbf{y}$, respectively. It is assumed that these errors are normal variables and are unbiased. The $\mathbf{B}$ matrix is thus related to the accidental errors of $\mathbf{x_b}$, while the model systematic errors or biases are assumed to be zero, as pointed out in [8]. Matrix $\mathbf{B}$ is defined as follows:

$$B \equiv E\left((x_b - x_t)\cdot(x_b - x_t)^T\right) \tag{2}$$

where $\mathbf{E}$ represents the expected value of the elements of the matrix $(x_b - x_t)\cdot(x_b - x_t)^T$, and $\mathbf{x_t}$ is the vector of the true values of the variables contained in $\mathbf{x_b}$. Note the actual values of $\mathbf{x_t}$ are unknown, so it becomes necessary to define a technique to estimate the values of $\mathbf{B}$ from available information.

$\mathbf{H(x)}$ is an operator that yields results analogous to the variables $\mathbf{y}$ from the model prognostic variables included in $\mathbf{x}$. As an example, if an element of $\mathbf{y}$ represents the measurement of the meridional wind at a certain geographical location and a certain height, $\mathbf{H(x)}$ interpolates the values of the meridional wind field obtained from the numerical model at the geographical location and the height of this measurement. The H operator is generally non-linear, but it is possible to linearize it with the following approximation:

$$H(x) = H(x_b + x - x_b) \approx H(x) + \nabla H(x - x_b), \tag{3}$$

where $\mathbf{\nabla H}$ is the gradient of $\mathbf{H(x)}$. If the linearized expression for $\mathbf{H(x)}$ is used, $\mathbf{J(x)}$ assumes a quadratic form.

It is assumed that the errors of the field measurements contained in vector $\mathbf{y}$ are statistically independent of each other, so $\mathbf{R}$ is a diagonal matrix; its diagonal contains the variances of the errors of each field measurement.

The matrix $\mathbf{B}$ is essential to this assimilation system. First, its non-diagonal terms contain the covariances of $\mathbf{x_b}$ errors at different grid points and also the covariances of errors of different variables. These covariances are necessary to propagate the information related to any field measurement through the horizontal and vertical directions and allow measurement of a specific variable to affect the analysis of others. In addition to this, the matrices $\mathbf{B^{-1}}$ and $\mathbf{R^{-1}}$ together determine the relative importance of the background conditions $\mathbf{x_b}$ and the field measurements $\mathbf{y}$ to determine the analysis $\mathbf{x_a}$.

The WRF-DA system offers two methods to estimate $\mathbf{B}$: the National Meteorological Center method (NMC), described by Parrish and Derber [15], and a method based on ensembles of predictions ([1], Chapter 9). Due to the availability of computer resources, we used the NMC method, which is relatively more economic. This method requires a database of pairs of $\mathbf{x_b}$; each pair has two short-term predictions for the same hour obtained with different forecast horizons (for example, 12 and 24 h), and consequently with different initial conditions. The pairs of predictions correspond to certain hours of the day (for example, 0:00 and 12:00 GMT), and cover a certain period (for example, one year). It is assumed that the differences between the forecasts of each pair have statistical properties analogous to those of the background condition errors, and in particular, to estimate B, the following approximation can be used:

$$B = E\left((x_b - x_t)^T \cdot (x_b - x_t)\right) \approx E\left(\left(x_b^{12hs} - x_b^{24hs}\right)^T \cdot \left(x_b^{12hs} - x_b^{24hs}\right)\right) \tag{4}$$

where x_b^{12hs} and x_b^{24hs} are a generic pair of predictions with 12 and 24 h time horizons, respectively. The validation of this technique to estimate **B** is empirical.

Note that, if **n** is the length of the $\mathbf{x_a}$, and $\mathbf{x_b}$ vectors, the size of **B** is **nxn**. This implies a matrix of very large dimensions that would cause important technical difficulties to operate with **B** or $\mathbf{B^{-1}}$, and even to store them in the computer memory. The 3D-Var technique implemented in the WRF-DA system solves this problem by factoring **B** as a product of certain matrices that have clear physical interpretations, allowing certain assumptions about their structures to be made and making them computationally tractable [2,4].

As a final remark in this subsection, we noted that both for the simulations of the background conditions and the estimation of the **B** matrix, a specific configuration of the WRF model was used, which here is the one described in Section 2.

WRF-DA Implementation for This Work

To apply the NMC method to estimate the matrix **B**, we used WRF-GFS simulations extended for 24 h and initialized at 0:00 GMT and 12:00 GMT during the entire year of 2016. The results obtained for 12 and 24 h after the initial conditions were used to estimate the matrix **B**, as indicated in the previous section. In order to gain insight into the spatial structure of the covariances contained in **B**, Decombes et al., [16] and Rivi [17] proposed "pseudo-single observation tests". Such tests consist of choosing a particular variable of $\mathbf{x_b}$ in a particular grid point and suppose a hypothetical observation that increments in a fixed amount the background value of this variable at the selected grid point. The data assimilation process is carried on considering this single pseudo observation and prescribing a hypothetical value of the standard deviation of its error. Rivi [17] showed that the $\mathbf{x_a} - \mathbf{x_b}$ difference was proportional to the covariance between the errors of the background variable in question at the chosen grid point and the errors of the rest of the background variables at all the grid points. The plots of the $\mathbf{x_a} - \mathbf{x_b}$ differences for selected variables illustrate how the **B** matrix spreads the information of field measurements. In Appendix B, we summarized the fundamentals of the pseudo observation tests, and we showed some selected results for the **B** matrix estimated here.

The background conditions $\mathbf{x_b}$ were obtained from WRF-GFS simulations initiated at 0:00, 6:00, 12:00, and 18:00 GMT, during those months of 2017 for which information from the WF1 and WF2 stations was available. We used the hourly results of these simulations from 4 to 9 h since the initialization of each simulation. In this way, four consecutive simulations could cover the 24 h of each day. The selected forecast horizon was the earliest for which the GFS prediction results were available in real-time, with some time left to carry out the processes described in this work. In this way, it is possible to implement these processes in an operative mode. For each local hour in Uruguay, Table 1 shows which WRF-GFS initialization cycle was used, and which hour within its forecast horizon corresponded to the background condition for that local hour.

Table 1. Hours within the WRF simulations used to define the background conditions as a function of the initialization cycle and local time in Uruguay. Each row corresponds to an initialization cycle, and each column corresponds to a local hour in Uruguay. The hours used within each cycle are indicated, considering the hours elapsed since its initialization.

Initialization Cycle	Local UY Hour																							
	0*	1	2	3	4	5	6	7	8	9	10	11	12	13	14	15	16	17	18	19	20	21	22	23
0:00		4	5	6	7	8	9																		
6:00								4	5	6	7	8	9											
12:00														4	5	6	7	8	9					
18:00	9*																			4	5	6	7	8

*: Local time 0 uses the results at the ninth hour of the simulation started at 18:00 GMT on the previous day as a background condition.

In the present study, the assimilation made for each hour was independent of the previous assimilations; the corresponding background condition only incorporated information from the GFS global simulation. This is called a "cold start". Alternatively, the background conditions could also have considered the field measurements made in previous hours. This alternative is called a warm start. The warm start method was tested in the context of the present work and obtained results of equivalent quality to those obtained with the cold start, so here we showed only the latter results. The similarity of the results from both methods might be because the assimilated information was very local; the assimilation of data in a wider region could influence the simulations in the area of interest over more hours and thus increase the relevance of assimilation from previous hours. The verification of this hypothesis requires a database of field measurements more complete than the one that was available for this work.

Each field measurement is specified to the WRF-DA system in a file that contains the day and time of the measurement, the geographic coordinates, and ground elevation of the corresponding station, the height of the measurement sensor, and the records of wind speed and direction. This information is used to generate the vector of field observations, y. Considering that the topography of the regional model has some degree of smoothing, the elevation of the specified terrain is not the actual elevation, but the elevation corresponding to the topography of the model interpolated to the station's location coordinates. The height specified for the sensors is the elevation of the station plus 100 m. The specified module of wind speed is the average of the anemometer record for two consecutive 15-min periods centered on the hour for which the assimilation will be performed.

The specified wind direction is deduced from the background condition, interpolating each component of the wind vector to the geographical position and elevation of the anemometer. Although vanes are available at the stations, it was found that using their records produces results that are not as satisfactory as those obtained by deducing the wind direction from the background condition. This suggests that the quality of data obtained from these weathervanes is relatively poor, while regional short-term predictions are reasonably good with regard to this variable. It may be of interest to evaluate the effect of the assimilation of atmospheric pressure measurements on the wind direction obtained in the analysis, but no such observations were available to complete this analysis.

In addition, to estimate the **R** matrix, the WRF-DA system uses a file that specifies the standard deviations of the errors of the different kinds of field observations (obserr.txt). In the present work, we adjusted the errors specified in this file, proposing a value of 0.1 m/s for wind speed, which is reasonable for the type of sensors installed in the stations considered here [18]. For wind direction, we kept the default value proposed in the obserr.txt file, 5°. We also pointed out that the WRF-DA system uses two "namelist" files to prescribe parameters to be set for the **B** matrix estimation and the computation of the resultant analysis (x_a vector). In this work, we used the default values for these parameters, which are indicated in the WRF-DA user guide [14].

4. Results

To evaluate the quality of the obtained pseudo-observations, we interpolated each component of the wind vector to the geographic location of VSt and the levels of 87 and 36 m above the ground, and then we computed the wind speed at these levels. Figure 4a shows the systematic error or bias of the two levels for each hour of the day. The bias is defined as the average of the wind estimate at a given level and hour minus the corresponding observed value over all the studied days:

$$b \equiv \overline{s_a} - \overline{s_{obs}} \tag{5}$$

where s_a is the wind speed interpolated from the analysis to the location and level of each Vst sensor, and s_{obs} is the correspondent measured wind speed. The overline represents the average over all the studied days. Figure 4b shows the relative bias, defined as the bias divided by the observed mean wind speed at the corresponding hour:

$$b_{rel} \equiv \frac{\overline{s_a} - \overline{s_{obs}}}{\overline{s_{obs}}} \tag{6}$$

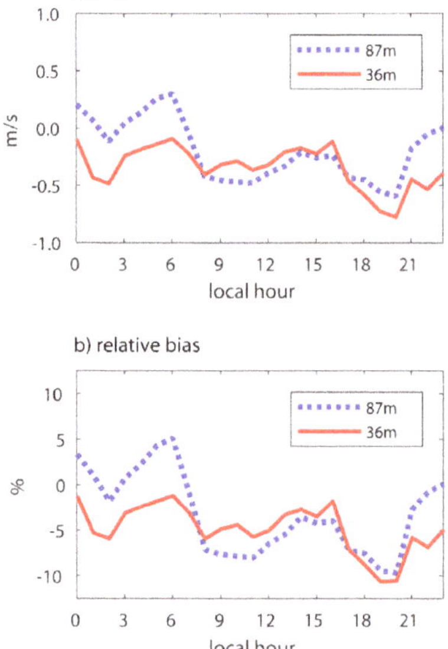

Figure 4. (**a**) Bias and (**b**) relative bias of wind estimates, which are 87 m above ground (red solid line) and 36 m above ground (blue dotted line), at VSt as a function of the local hour.

At 87 m, the bias was moderate, generally smaller than 0.5 m/s, while relative bias was generally smaller than 5%. At 36 m, the bias and the relative bias were similarly moderate: the bias was generally smaller than 0.5 m/s, while the relative bias during some hours was slightly larger than that found at 87 m. Note that 87 m above the ground was similar to the elevation of the assimilated observations, while the 36-m elevation was significantly closer to the ground. The moderate systematic errors found at both elevations indicated that the data assimilation technique effectively combined the information from short-term WRF predictions and wind measurements. The wind measurements from the wind farms lacked information about elevations relatively close to the ground, e.g., at 36 m, while the WRF simulations did include this information since they considered several elevations within the first 100 m above the ground but had errors of their own. The assimilation process corrected these errors at the locations and elevations of the anemometers in the wind farms and also transmitted the effects of these gains to other regions and elevations. The information that allowed these gains to spread was contained in the background error matrix **B**. To further illustrate this point, Figure 5 shows the bias of the background conditions 36 m above the ground at VSt and the bias resulting from the assimilation process (also included in Figure 4a). The background bias was larger for all hours. Since both biases were means of errors, it was possible to compute the significance of their difference with a two-tailed Student t-test [19]. It was found that these biases were different with a significance value lower than 0.05 for the local hours from 0:00 to 16:00, and from 21:00 to 23:00 (the hours with significant differences are indicated with a green bar in Figure 5).

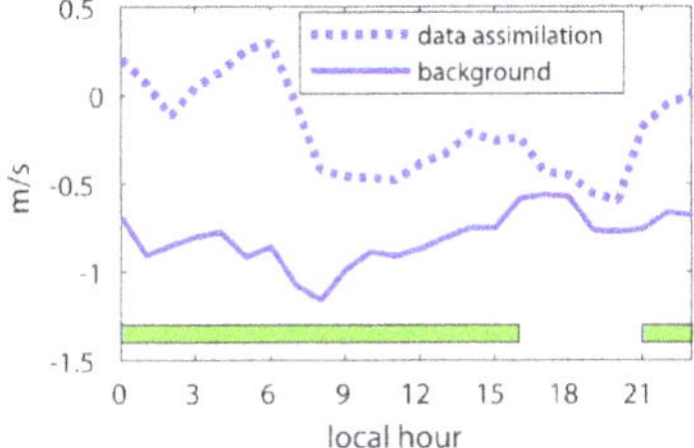

Figure 5. The bias of two wind estimates 36 m above ground at VSt, as a function of the local hour. The solid line indicates the wind estimates from the background conditions, and the dotted line indicates the wind estimates from the assimilation process. The green bar indicates the hours for which the difference between both biases was significant at the 5% level, computed as indicated in the text.

Next, we presented two statistical parameters related to accidental errors. Figure 6 shows the Pearson correlation of estimated versus observed wind speeds at the VSt station. At 87 m above the ground, there were correlations generally larger than 0.75 during the night hours, and generally larger than 0.80 during the day. At 36 m, the values were slightly smaller at nighttime and very similar during the day. Figure 7 shows the standard deviation of the error (estimated minus observed values) divided by the mean observed wind speed at each hour (RSTD). RSTD was approximately 25% for both elevations considered.

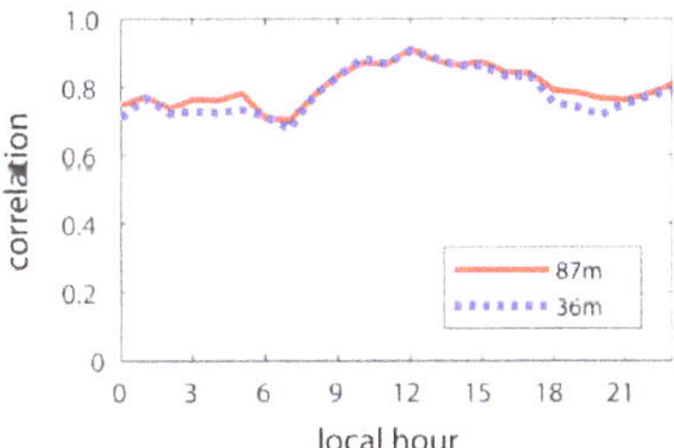

Figure 6. Pearson correlation of estimated wind versus observed wind as a function of the local hour at VSt. The red solid line shows the correlation at 87 m above the ground, and the blue dotted line shows at 36 m.

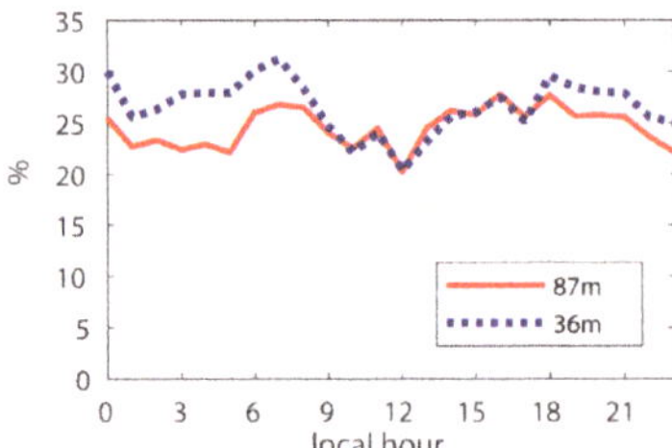

Figure 7. The relative standard deviation of wind estimate error as a function of the local hour at VSt. The red solid line shows the correlation at 87 m above the ground, and the blue dotted line shows at 36 m.

Figure 8 shows scatter plots of estimated versus observed wind speeds at 7:00 and 12:00 and 87 and 36 m above the ground. These hours were chosen because they represent the periods of the day with the relatively worst and best adjustments, as assessed in Figures 6 and 7. Samples of assimilated values and the corresponding observed wind speed values, including those shown in Figure 8, made it possible to estimate probability distributions of the errors, such as empirical percentiles, from which confidence intervals for the wind speed estimates could be calculated.

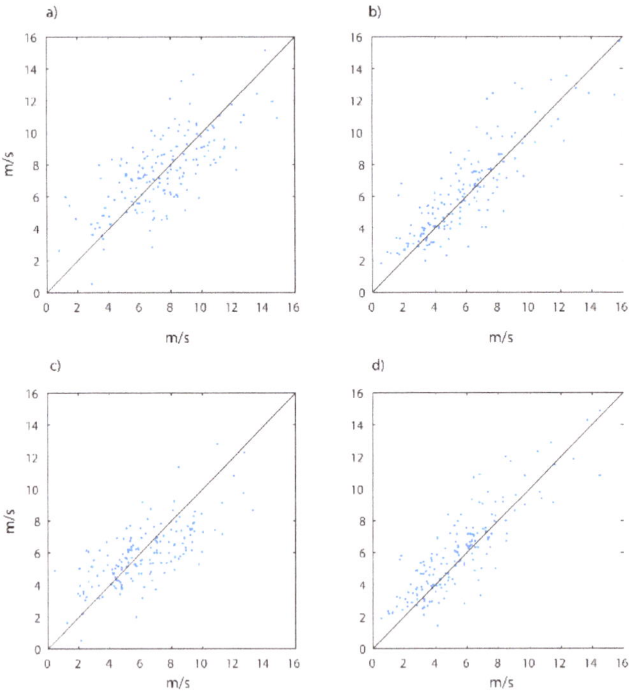

Figure 8. Estimated wind versus observed wind 100 m above the ground at VSt at (**a**) 7:00 and (**b**) 12:00. Estimated wind versus observed wind 30 m above ground at VSt at (**c**) 7:00 and (**d**) 12:00.

5. Summary and Conclusions

This work assessed the quality of wind speed estimates in Uruguay obtained with the WRF-DA system, which was used to assimilate wind speed measurements 100 m above the ground at two wind farms. The quality of the estimates was assessed with an anemometric station placed between the wind farms, that measured wind speed at 87 and 36 m above the ground. The information to be assimilated from field measurements was minimal, not only because it included only two stations but also because it lacked records of other atmospheric variables that are related to wind, such as atmospheric pressure. It was interesting to assess the extent to which these minimum field measurements could generate useful interpolations and evaluate the quality of the wind estimates at elevations both similar and different to those of the assimilated records. The measurement station used to validate the wind speed assessments was placed between those used for the assimilation process, so the conclusions of the assessment are applicable only to regions between the two main stations.

Wind speed estimates showed a low systematic error at the verification station, generally below 0.5 m/s at both 87 and 36 m above the ground. A relative systematic error was generally less than 5% of the average speed. This result indicated that the data assimilation technique effectively combined information from field measurements and background conditions. The assimilated measurements did not include information from elevations as low as 36 m. The background conditions did contain information from these low elevations, but with systematic errors of their own. The assimilation technique managed to propagate the gain from the observations at 100 m above the ground in the wind farms to other regions and to lower elevations. The covariance matrix of the background condition error was essential to the propagation of these observational gains.

As for the total error, the correlation values between observed and estimated wind speed and the standard deviation of the total error of each estimate, generally about 1 m/s to 1.5 m/s, suggested that the obtained estimates could be of sufficient quality to be useful in various applications. Some examples of applications in which such estimates are valuable are the estimation of wind climatology within the range of the considered height levels, retrospective simulations of transport processes and dispersion of air pollutants, or real-time estimation of environmental conditions in which systems whose operation can be affected by the wind are being used. In any case, the effective use of pseudo-observations in a specific application requires the estimation of their confidence intervals, which are necessary both to assess whether the accuracy of pseudo-observations is acceptable for the application in question and to take into account the effects of the uncertainty of these data if they are used. The generation of databases of pseudo-retrospective observations, such as the one presented in this work, allows for the estimation of these confidence intervals.

For future studies, we are interested in quantifying the effects of including atmospheric pressure observations on the quality of the results. We are also interested in evaluating the effect of expanding the region in which observations are collected for the assimilation of data on the results of the hot start option.

Finally, we pointed out that the topography of the studied region is not completely flat but is relatively smooth. This can contribute to the quality of results obtained by assimilating a few observations in numerical simulations with relatively low resolution. In the case of regions with relatively complex topography, the numerical simulation may require finer spatial resolution to properly take into account the effects of topography on the wind field. The proper quantity and location of measurement stations should also be evaluated in each case.

Author Contributions: Conceptualization, G.C.B., S.O.; methodology, G.C.B., S.O.; software, G.C.B., S.O.; validation, G.C.B., S.O.; formal analysis, G.C.B., S.O.; investigation, G.C.B., S.O.; resources, G.C.B.; data curation, G.C.B. and S.O.; writing—original draft G.C.B.; visualization, G.C.B.; supervision, G.C.B.; project administration, G.C.B.; funding acquisition, G.C.B.

Funding: This research was funded by UTE and the Comisión Sectorial de Investigación Científica (CSIC) of the Universidad de la República, Uruguay, grant number CSIC-UTE-2017-33. The Agencia Nacional de Investigaciòn Científica, ANII, Uruguay, founded earlier stages of this work, grant FSE-2011-6562.

Acknowledgments: UTE provided the data to do this study. The reviewers of this work made valuable comments and indications that contributed to its improvement.

Conflicts of Interest: The authors declare no conflicts of interest.

Appendix A

The setting of the WRF model used in this work is analogous to that of the work by Cazes Boezio and Ortelli [12] which evaluates short-term forecasts of wind power generated in Uruguay. The horizontal resolution was 30 km in the zonal and meridional direction. Figure 3 shows the grid points used to compute air temperature, pressure, density, and vertical velocity. The grid points used to compute the zonal and meridional velocities (not shown) were staggered one half of the horizontal resolution in the zonal and the meridional directions, respectively, according to the Arakawa C-grid arrange [1].

The model setting employed considered 54 layers in the vertical direction. The model vertical coordinate is η [1], defined as

$$\eta = \frac{p_d - p_T}{p_S - p_T} \tag{A1}$$

where p_d is the hydrostatic component of dry air pressure at a particular atmosphere level, and p_S and p_T are the analogous pressures at the Earth's surface and the atmosphere conventional top, respectively. In this work, the atmosphere top was set to 50 hPa. Table A1 gives the values of h at each layer interface.

Table A1. η values at the top of each layer of the vertical discretization.

Layer Number	h Value at Layer Top	Layer Number	h Value at Layer Top	Layer Number	h Value at Layer Top
1	0.9880	19	0.9240	37	0.7135
2	0.9969	20	0.9165	38	0.6911
3	0.9950	21	0.9088	29	0.6668
4	0.9935	22	0.9008	40	0.6406
5	0.9935	23	0.8925	41	0.6123
6	09910	24	0.8840	42	0.5806
7	0.9899	25	0.8752	43	0.5452
8	0.9861	26	0.8661	44	0.5060
9	0.9821	27	0.8567	45	0.4630
10	0.9777	28	0.8471	46	0.4161
11	0.9731	29	0.8371	47	0.3656
12	0.9682	30	0.8261	48	0.3119
13	0.9629	31	0.8141	49	0.2558
14	0.9573	32	0.8008	50	0.1982
15	0.9513	33	0.7863	51	0.1339
16	0.9450	34	0.7704	52	0.0804
17	0.9382	35	0.7531	53	0.0362
18	0.9312	36	0.7341	54	0.0000

The horizontal velocities and the air temperature were computed inside each layer, while the vertical velocity was computed at the layer interfaces, according to the Lorenz vertical grid arrangement [1].

The WRF model allowed us to choose several parameterizations of physical processes, especially for atmospheric boundary layer processes, surface layer processes, short and long wave radiative heat transfers, convective precipitation, clouds microphysics, and drag associated with gravity waves. Table A2 shows the parameterizations chosen in the current work and indicates references for them.

Table A2. Parameterization of physical processes used.

Physical Process	Scheme Used
Short Wave Radiation	Dudhia scheme [20]
Long Wave Radiation	RRTM schene [21]
Surface Layer	Revised MM5 surface layer scheme [22]
Atmospheric Boundary Layer	Yonsei University scheme [23]
Microphysics	Hong et al. scheme [24]
Cumulus Precipitation	Simplified Arakawa Schubert scheme [25]
Gravity Wave Drag	Kim Arakawa scheme [26]
Land Processes	Noah land surface model [27]

Appendix B

Kalnay [3] showed the result of the optimization indicated in Equation (A1), that yields the 3D-Var analysis $\mathbf{x_a}$ and is equivalent to the result of the optimal interpolation procedure,

$$x_a - x_b = B(\nabla H)^T \cdot [\nabla H \cdot B \cdot (\nabla H)^T + R]^{-1} \cdot [y - (\nabla H)(x_b)] \tag{A2}$$

We chose a particular variable and a particular grid point that corresponds to the kth element of $\mathbf{x_b}$ or $\mathbf{x_a}$, according to the conventional order of these vectors. We defined the synthetic observation $\mathbf{y}$ at the chosen grid point as the kth value of $\mathbf{x_b}$ plus a conventional increment Δ,

$$y = x_b(k) + \Delta \tag{A3}$$

Since the **∇H** operator produces analogous variables to those of the vector of observations $\mathbf{y}$ from $\mathbf{x_b}$, and the analogous to the synthetic observation $\mathbf{y}$ in the background condition is $\mathbf{x_b}(k)$, the correspondent **∇H** is a vector with all its terms equal to 0, except the kth element, which is equal to 1, so

$$H(x_b) = x_b(k), \text{ and } y - H(x_b) = \Delta \tag{A4}$$

The matrix **R** of covariance of observation errors has a single element, which is the covariance of the synthetic observation $\mathbf{y}$. We chose a conventional value s^2 for this covariance. Rizvi [17] showed that with these choices, Equation (A2) yields

$$x_a - x_b = \frac{B_k}{b_{kk} + \sigma^2} \tag{A5}$$

where $\mathbf{B_k}$ is the k column of **B**. Equation A5 indicates that the $\mathbf{x_a} - \mathbf{x_b}$ difference is proportional to a vector that gives all the covariances of the error of $\mathbf{x_b}(k)$ with the errors of all the other variables of $\mathbf{x_b}$. Note that the $\mathbf{x_a} - \mathbf{x_b}$ difference is independent of the $\mathbf{x_b}$ condition chosen. Plots of $\mathbf{x_a} - \mathbf{x_b}$ fields for a selected variable at a selected level of the numerical domain can help to understand the geographical structure of the error covariances of that variable at that level.

Here, we chose to increase in 1 m/s the zonal ("test A") and the meridional ("test B") wind of a particular $\mathbf{x_b}$ condition at the grid point and level closest to the location of the Vst station and the level of 87 m above ground. s was chosen as 1 m/s. Figure A1a shows the $\mathbf{x_a} - \mathbf{x_b}$ field for zonal velocity resulting from test A, at the same model level of the selected grid point (level 7 from the ground). Figure A1b shows the $\mathbf{x_a} - \mathbf{x_b}$ field for the meridional velocity at the same level, for "test B". These results indicated that WF1 and WF2 were located in regions, where the zonal and meridional winds are well correlated with those of the Vst, and, therefore, wind estimates at this location benefit from observational gains obtained at WF1 and WF2. Besides, this type of analysis can be useful to define the density of a station network intended to cover a specific area.

Figure A2 shows the vertical profile of the $x_a - x_b$ zonal wind difference for test A, at the point of the horizontal grid in which the increase of zonal wind speed was prescribed. This profile shows the vertical structure of the covariances of zonal velocity background errors with the background error at the level chosen for the test. It was found that these covariances increased with the elevation up to 500 m above the ground, and then decreased to values that were close to 0 in the upper atmosphere (Figure A2a). Figure A2b shows a zoom of the profile shown in Figure A2a for the first 150 m above the ground, in order to focus on the levels of interest to this work. Although covariances were lower at lower elevations, their relatively large values indicated that observational gains obtained at elevations about 100 m could propagate quite directly to lower levels. The analogous vertical profiles from test B were found to be very similar to those from test A, and are not shown here.

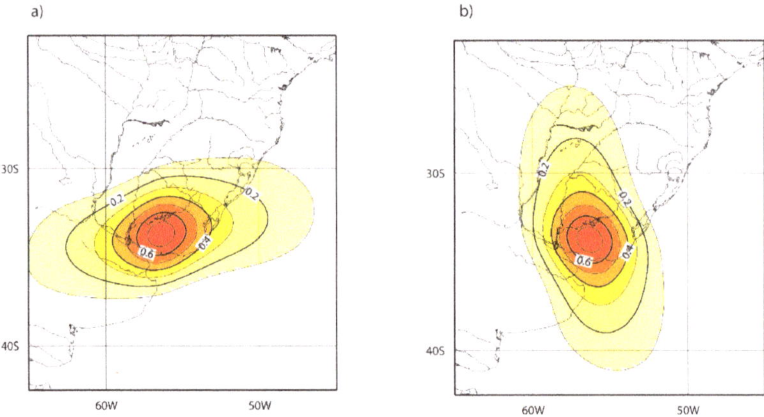

Figure A1. (a) $x_a - x_b$ zonal wind difference in test A at the 7th layer of the model; (b) $x_a - x_b$ meridional wind difference in test B at the same level. Contour interval, 0.1 (dimensionless).

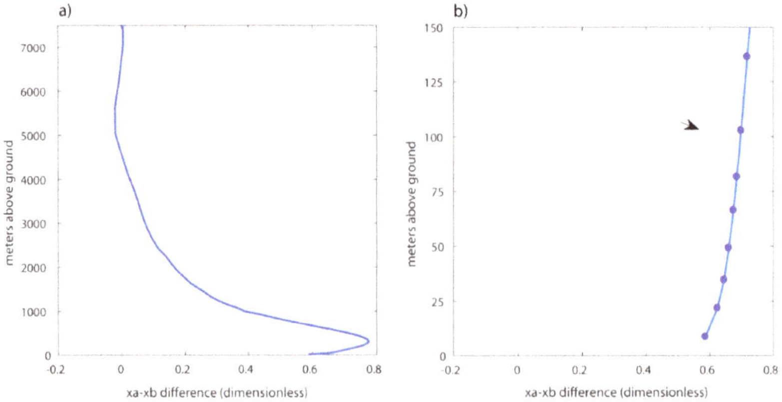

Figure A2. (a) Vertical profile of $x_a - x_b$ zonal wind differences in test A at the horizontal grid point closest to Vst location (for the first 8000 m above the ground). Abscissa, $x_a - x_b$ (dimensionless), ordinate, elevation above ground; (b) zoom of Figure A2, a vertical profile at the first 150 m above the ground. Dots indicate the model grid points, and the arrow, the grid point at which the perturbation is prescribed.

References

1. Skamarock, W.C.; Klemp, J.B.; Dudhia, J.; Gill, D.O.; Barker, D.M.; Wang, W.; Powers, J.G. A Description of the Advanced Research WRF Version 3. NCAR Tech Note NCAR/TN-475+STR. Available online: http://citeseerx.ist.psu.edu/viewdoc/summary?doi=10.1.1.484.3656 (accessed on 25 October 2019).
2. Barker, D.; Huang, X.Y.; Liu, Z.; Auligné, T.; Zhang, X.; Rugg, S.; Ajjaji, R.; Bourgeois, A.; Bray, J.; Chen, Y.; et al. The Weather Research and Forecasting Model's Community Variational/Ensemble Data Assimilation System, WRFDA. Available online: https://journals.ametsoc.org/doi/pdf/10.1175/BAMS-D-11-00167.1 (accessed on 25 October 2019).
3. Kalnay, E. *Atmospheric Modelling, Data Assimilation and Predictability*; Cambridge University Press: Cambridge, UK, 2003; p. 345.
4. Barker, D.; Huang, X.Y.; Guo, Y.-R.; Xiao, Q.N. A three dimensional (3Dvar) data assimilation system for use with MM5. Implementation and Initial Results. *Mon. Weather Rev.* **2004**, *132*, 897–914. [CrossRef]
5. Huang, X.Y.; Xiao, Q.; Barker, D.M.; Zhang, X.; Michalakes, J.; Huang, W.; Henderson, T.; Bray, J.; Chen, Y.; Ma, Z. Four-dimensional variational data assimilation for WRF: Formulation and preliminary results. *Mon. Weather Rev.* **2009**, *137*, 299–314. [CrossRef]
6. Wang, X.; Barker, D.M.; Snyder, C.; Hamill, T.M. A hybrid ETKF-3DVAR data assimilation scheme for the WRF model. Part I: Observing system simulation experiment. *Mon. Weather Rev.* **2008**, *136*, 5116–5131. [CrossRef]
7. Wang, X.; Barker, D.M.; Snyder, C.; Hamill, T.M. A hybrid ETKF-3DVAR data assimilation scheme for the WRF model. Part II: Real observation experiments. *Mon. Weather Rev.* **2008**, *136*, 5132–5147. [CrossRef]
8. Harlim, J. Model Error in Data Assimilation. In *Nonlinear and Stochastic Climate Dynamics*; Franzke, C., O'Kane, T., Eds.; Cambridge University Press: Cambridge, UK, 2017; pp. 276–317. [CrossRef]
9. Rao, V.; Sandu, S. A-posteriori error estimates for inverse problems. *Procedia Comput. Sci. J.* **2018**, *20*, 1256–1265.
10. Orteli, S.; Boezio, G.C. Construction of Empirical Speed-Power Curves in Wind Farms Installed in Uruguay. Application to Real-Time Data Quality Control and Estimation of Potential Generation in Case of Restrictions. X Brazilian Micrometeorology Workshop. Available online: https://www.even3.com.br/anais/micrometeorologia2017/64488-construcao-de-curvas-de-velocidade-potencia-empiricas-em-parques-eolicos-instalados-no-uruguai-aplicacao-ao-contr/ (accessed on 25 October 2019).
11. Cornalino, E.; Draper, M. Planning the Distribution of Wind Farms in Uruguay in Order to Optimize the Operability of Large Amounts of Wind Power. Available online: www.fing.edu.uy/cluster/eolica/publi/Cornalino_Draper_EWEC2012_FINAL.pdf (accessed on 29 October 2019).
12. Cazes Boezio, G.; Orteli, S. Minimum-cost numerical prediction system for wind power in Uruguay, with an assessment of the diurnal and seasonal cycles of its quality. *Sci. Nat.* **2018**, *40*, 206–210.
13. Barker, D.; Huang, W.; Guo, Y.-R.; Bourgeois, A. A Three-Demiensional Variational (3DVAR) Data Assimilation System for Use with MM5 (No. NCAR/TN-453+STR). Available online: https://opensky.ucar.edu/islandora/object/technotes%3A309 (accessed on 25 October 2019).
14. User's Guide for the Advanced Research WRF (ARW) Modeling System Version 3.9 WRF-ARW, Chapter 6. Available online: http://www2.mmm.ucar.edu/wrf/users/wpsv3.9/known-prob-3.9.html (accessed on 25 October 2019).
15. Parrish, D.F.; Derber, J.C. The National Meteorological Center's spectral statistical interpolation analysis system. *Mon. Weather Rev.* **1992**, *120*, 1747–1763. [CrossRef]
16. Descombes, G.; Auligné, T.; Vandenberghe, F.; Barker, D.; Barré, J. Generalized background error covariance matrix model (GEN_BE v2.0). *Geosci. Model Dev.* **2015**, *8*, 669–696. [CrossRef]
17. Rizvi, S. WRF-DA Background Error (Modeling and Estimation), WREF-DA Tutorial Presentation Slides, Chapter 4. University Corporation for Atmospheric Research, 2016. Available online: www2.mmm.ucar.edu/wrf/users/wrfda/Tutorials/2016_Aug/docs/WRFDA_BE.pdf (accessed on 29 October 2019).
18. Cataldo, J.; Universidad de la República, Montevideo, Uruguay. Personal communication, 2019.
19. Press, W.H.; Taukolsky, S.A.; Vettering, W.T.; Flannery, B.P. *Numerical Recipes in Fortran 77: The Art of Scientific Computing*; Cambridge University Press: Cambridge, UK, 1996; p. 1010.

20. Dudhia, J. Numerical study of convection observed during the winter monsoon experiment using a mesoscale two-dimensional model. *J. Atmos. Sci.* **1989**, *46*, 307–310. [CrossRef]
21. Iacono, M.J.; Delamere, J.S.; Mlawer, E.J.; Shephard, M.W.; Clough, S.A.; Collins, W.D. Radiative forcing by long-lived greenhouse gases: Calculations with the AER radiative transfer models. *J. Geophys. Res.* **2008**, *113*, D13103. [CrossRef]
22. Jimenez, P.; Dudhia, J.; Gonzalez-Ruoco, J.F.; Navarro, J.; Montavez, J.P.; Garcia-Bustamente, F. A revised scheme for the WRF surface layer formulation. *Mon. Weather Rev.* **2012**, *140*, 898–918. [CrossRef]
23. Hong, S.-Y.; Noh, Y.; Dudhia, J. A new vertical diffusion package with an explicit treatment of entrainment processes. *Mon. Weather Rev.* **2006**, *134*, 2318–2341. [CrossRef]
24. Hong, S.Y.; Dudhia, J.; Chen, S.H. A Revised Approach to Ice Microphysical Processes for the Bulk Parameterization of Clouds and Precipitation. *Mon. Weather Rev.* **2004**, *132*, 103–120. [CrossRef]
25. Grell, G.A. Prognostic evaluation of assumptions used by cumulus parameterizations. *Mon. Weather Rev.* **1993**, *121*, 764–787. [CrossRef]
26. Kim, Y.J.; Arakawa, A. Improvement of Orographic Gravity Wave Parameterization Using a Mesoscale Gravity Wave Model. *J. Atmos. Sci.* **1995**, *52*, 1875–1902. [CrossRef]
27. Niu, G.-Y.; Yang, Z.L.; Mitchel, K.E.; Chen, F.; Ek, M.B.; Barlage, M.; Kumar, A.; Manning, K.; Niogi, D.; Rosero, E. The community Noah land surface model with multiparameterization options (Noah-MP): 1. Model description and evaluation with local-scale measurements. *J. Geophys. Res.* **2011**, *116*, D12. [CrossRef]

Data Descriptor

A High-Resolution Global Gridded Historical Dataset of Climate Extreme Indices

Malcolm N. Mistry [1,2]

[1] Department of Economics, Ca' Foscari University of Venice, 30121 Venice, Italy; malcolm.mistry@unive.it
[2] Centro Euro-Mediterraneo sui Cambiamenti Climatici (CMCC), 30175 Venice, Italy

Received: 18 February 2019; Accepted: 9 March 2019; Published: 13 March 2019

Abstract: Climate extreme indices (CEIs) are important metrics that not only assist in the analysis of regional and global extremes in meteorological events, but also aid climate modellers and policymakers in the assessment of sectoral impacts. Global high-spatial-resolution CEI datasets derived from quality-controlled historical observations, or reanalysis data products are scarce. This study introduces a new high-resolution global gridded dataset of CEIs based on sub-daily temperature and precipitation data from the Global Land Data Assimilation System (GLDAS). The dataset called "CEI_0p25_1970_2016" includes 71 annual (and in some cases monthly) CEIs at $0.25° \times 0.25°$ gridded resolution, covering 47 years over the period 1970–2016. The data of individual indices are publicly available for download in the commonly used Network Common Data Form 4 (NetCDF4) format. Potential applications of CEI_0p25_1970_2016 presented here include the assessment of sectoral impacts (e.g., Agriculture, Health, Energy, and Hydrology), as well as the identification of hot spots (clusters) showing similar historical spatial patterns of high/low temperature and precipitation extremes. CEI_0p25_1970_2016 fills gaps in existing CEI datasets by encompassing not only more indices, but also by being the only comprehensive global gridded CEI data available at high spatial resolution.

Dataset: https://doi.org/10.1594/PANGAEA.898014

Dataset License: CC-BY: Creative Commons Attribution 4.0 International

Keywords: climate extreme indices (CEIs); ClimPACT; GLDAS; Expert Team on Climate Change Detection and Indices (ETCCDI); Expert Team on Sector-specific Climate Indices (ET-SCI)

---

1. Introduction

Extremes in climate such as floods, droughts, and cold and heat-waves can have significant societal, ecological, and economic impacts globally [1]. Since the publication of the third assessment report of the Intergovernmental Panel on Climate Change (IPCC) in 2000, characterizing extremes under past and projected future climate has generated rapid interest [2]. The climate modelling community, for instance, has spent increasing effort to capture high-frequency extreme events in their simulations of historical and future projected climate. The underlying aim for both regional and global climate modelling exercises (e.g., CORDEX and PRIMAVERA)[1] has been to develop a better understanding of the evolution of extreme weather events under long-term climate change and variability.

The impetus to better understand extreme weather events is further driven by the impact modellers who assess sectoral damages at varying spatial scales. The two vital characteristics of

---

[1] CORDEX: http://www.cordex.org/; PRIMAVERA: https://www.primavera-h2020.eu/.

climate that are at the core of impact models are (i) mean climate and (ii) the occurrence and frequency of extreme events [3]. An increasing notion shared within the climate research community is that even a relatively small change in the frequency or severity of extreme weather events (i.e., in the tails of the probability distribution function) would have profound impacts on life and assets [4], thus making it further imperative to analyze extremes at higher temporal and spatial resolutions. For the scientific community focusing on impacts of climate change and variability, historical observations of extreme indicators can facilitate a better understanding of the role of extreme events and sectoral implications [5].

Largely driven by the requirement for a robust definition of climate extreme indicators, the Expert Team on Climate Change Detection and Indices (ETCCDI)[2] in 1999 led the first efforts in defining a set of climate extreme indices (CEIs) that provide a comprehensive overview of temperature and precipitation statistics [4,6–8]. The ETCCDI has developed an internationally coordinated set of core climate indices consisting of 27 descriptive indices for moderate weather extremes[3] [9–11]. The preliminary set of these 27 core indices were drawn up keeping the detection and attribution needs of the research community in mind [10,11]. Noting the limitations of the ETCCDI indices with regard to restricted scope/usage in assessing sectoral impacts, additional sector-relevant indices were recommended and developed by the Expert Team on Sector-specific Climate Indices (ET-SCI) [9].

This study introduces a new open-access high-resolution global gridded ($0.25° \times 0.25°$)[4] dataset of 71 CEIs (including the original 27 ETCCDI indices), covering the period 1970–2016. The dataset (hereafter referred to as "CEI_0p25_1970_2016") aims to contribute to the existing CEI databases by making available the first comprehensive CEI dataset currently unavailable for the climate community at a high resolution with worldwide coverage. Moreover, a consistent global CEI dataset covering a long historical time period can lay a framework for not only analyzing observed changes in extremes, but also potentially improving information services on extremes at regional scales [10].

The CEI_0p25_1970_2016 are a set of core (Table S1 in Supplementary Materials) and non-core (Table S2 in Supplementary Materials) indices[5] as defined and developed by the ETCCDI/ET-SCI, and adopted by the World Meteorological Organization (WMO). The set of "core indices" refers to indices that were developed by ETCCDI targeting the research community focusing on "detection and attribution" in climate science (details in Section 4).

The rest of the paper is organized as follows. Section 2 describes the CEI_0p25_1970_2016 in detail. Section 3 discusses the underlying meteorological dataset and the tools/methodology used in the preparation of the CEI_0p25_1970_2016. Section 4 outlines the novelty, potential scope, application, and limitations of the CEI_0p25_1970_2016. Dataset availability, ongoing work, and some recommendations for future research are summarized in Section 5.

2. Dataset Description

2.1. Spatial and Temporal coverage of CEI_0p25_1970_2016

The CEIs included in this study encompass all but two indices[6] that are part of the complete list of 73 ETCCDI/ET-SCI core and non-core indices [9]. The CEI_0p25_1970_2016 is derived using meteorological variables from the reanalysis data product Global Land Data Assimilation

---

2   Formed by the World Meteorological Organization (WMO) Commission for Climatology (CCl).

3   Extreme events that by definition typically occur a few times annually rather than severe impact, decadal weather events. The indices for moderate weather extremes use absolute or percentile thresholds generally set at moderate values (e.g., 25 °C, 90th percentile).

4   ~27 km × 27 km at the equator.

5   https://www.wcrp-climate.org/data-etccdi.

6   The two indices Cooling and Heating Degree Days (CDD and HDD) are computed separately as part of another dataset of additional indices relevant for health and energy sectors, currently under preparation [12]. Further details are provided in Section 5.2.

System (GLDAS) [13]. GLDAS is a new generation of reanalysis developed jointly by the National Aeronautics and Space Administration (NASA) Goddard Space Flight Center (GSFC) and National Centers for Environmental Prediction (NCEP) [14]. Because the spatial extent of GLDAS covers all land north of 60° S, the indices in CEI_0p25_1970_2016 are also computed over the corresponding 1440 (longitude) × 600 (latitude) grid cells. Further description of GLDAS as well as the reasons for using it as a data source in this study are discussed in Section 3.

2.2. Other Existing Datasets Incorporating CEIs

While other similar historical gridded CEI datasets do exist, they are either (i) regional in coverage, (ii) at coarser resolution, or (iii) limited in the number of indices available for research purpose. Examples include (i) the 30 CEIs made available by E-OBS at 0.10° gridded resolution for Europe http://surfobs.climate.copernicus.eu/dataaccess/access_eobs_indices.php, (ii) the global 0.50° gridded resolution S-14 indices dataset of 27 core ETCDDI indices available at http://h08.nies.go.jp/s14/ [15], and (iii) the global 3.75° × 2.5° resolution HadEX2 and GHCNDex datasets of 27 core ETCDDI indices available at https://www.climdex.org/learn/datasets/ [6,7]. To the best of the author's knowledge, the present database CEI_0p25_1970_2016 is currently the only comprehensive high-resolution global-gridded historical dataset of ETCCDI/ET-SCI core and non-core indices.

3. Materials and Methods

3.1. Data Acquisition and Processing

The CEIs used in this study were computed utilizing the WMO ET-SCI recommended and developed R-software package "ClimPACT2"[7] [9]. R [16] is an open-source language and software environment, developed primarily (but not solely) for statistical computing, and is applied widely in climate research. Moreover, ClimPACT2 also makes use of several R subroutines, such as SPEI [17], and is designed for operating on parallel computing infrastructure.

For the computation of CEI, ClimPACT2 requires the following meteorological variables (i) maximum near-surface air temperature (TX), (ii) minimum near-surface air temperature (TN), and (iii) near-surface total precipitation (PR), all at daily timesteps. These variables in the native Network Common Data Form 4 (NetCDF4)[8] format were obtained from the GLDAS-version 2[9] [13,18,19], available at 3-hourly timesteps and a fine spatial resolution of 0.25° × 0.25°. GLDAS is a global high-resolution reanalysis dataset that incorporates satellite and ground-based observations, producing optimal fields of land surface states and fluxes in near real time [13].

For the purpose of computing the CEI, the 3-hourly gridded variables (TX, TN, and PR) were first temporally aggregated to construct daily mean TX and TN, and daily total PR, using a suite of command line operators from NetCDF Command Operators (NCO ver 4.3.4)[10] and Climate Data Operators (CDO ver 1.9.0)[11]. Indices based on percentile thresholds (e.g., WSDI and CSDI in Table S1) were computed using years 1970–2000 as the baseline period. For details on classification of CEIs (namely "percentiles", "absolute", "threshold", "duration", and "others"), readers are guided for further reading in [6–9].

---

[7] R version 3.5.0 ("Joy in Playing") x86_64 on Linux Centos 6.6 software architecture. ClimPACT2 was accessed on 23 September 2018 from https://github.com/ARCCSS-extremes/climpact2.

[8] NetCDF is a set of scientific software libraries, with self-describing and machine-independent data format. https://www.unidata.ucar.edu/software/netcdf/docs/.

[9] Data accessed from https://disc.gsfc.nasa.gov/ on 12 July 2018.

[10] NCO [20]: accessed on 14 July 2018 from http://nco.sourceforge.net/.

[11] CDO [21] accessed on 14 July 2018 from http://www.mpimet.mpg.de/cdo.

3.2. Choice of GLDAS as a Reanalysis Dataset for the Computation of CEIs

Vis-à-vis other global gridded reanalysis datasets, GLDAS offers several advantages. First, GLDAS provides a consistent quality-controlled long global gridded time-series of the required variables (i.e., TX, TN, and PR) at a high spatial resolution. Other reanalysis data products available were found to have either a coarser spatial resolution (e.g., ECMWF-ERA40 and JRA-55, both available from the mid-1950s but at 1.125°), or a shorter time series (e.g., newly released ECMWF-ERA5 at 0.281° from 1979–present day, and NCEP-CFSv2 at 0.205° from 2011–present day). Second, GLDAS runs in near-real-time, offering the potential to regularly update the database presented here.

The choice of GLDAS for computing the current set of indices was further motivated by its large number of additional meteorological (e.g., specific humidity, surface pressure), land surface state (e.g., soil moisture, surface temperature), and flux (e.g., evaporation, sensible heat flux) variables, not commonly available in other reanalysis data products for a long time-series and at a high spatial resolution[12]. While none of these additional variables are required for computing the current set of indices, another dataset [12] of sectoral indices that are not presently implemented in the ETCCDI/ET-SCI indices requires a subset of these variables (details in Section 5.2). The two datasets of indices (current and [12] under prep.) will together comprise a large (~85) number of indices both based on the same underlying GLDAS data, thus enabling the climate impacts community to access "ready-to-use" multi-sectoral indices.

GLDAS has been comprehensively evaluated using different regional/global reference datasets in earlier studies (e.g., see [14] who compare the GLDAS daily surface air temperature at 0.25° gridded resolution with two reference datasets): (a) Daymet data (2002 and 2010) for the conterminous United States at 1-km gridded resolution, and (b) global meteorological observations (2000) from the Global Historical Climatology Network (GHCN).

Equally well-documented are certain known limitations of the temperature and precipitation estimates in GLDAS. Whereas spatial details in high mountainous areas are not sufficiently estimated by the GLDAS data, the surface air temperature estimates are generally accurate, with some caution recommended for mountainous areas [14]. Previous studies that have incorporated GLDAS data include (i) [22] for impact assessment studies in energy sector, and (ii) [23,24] for the analysis of regional environmental conditions and changes. For a comprehensive list of GLDAS-related references, readers are referred to https://ldas.gsfc.nasa.gov/gldas/GLDASpublications.php.

4. Key Features, Scope of Application, and Limitations of CEI_0p25_1970_2016

4.1. Novelty of CEI_0p25_1970_2016

The CEI_0p25_1970_2016 is currently the only dataset providing researchers and policymakers with an exhaustive list of ETCCDI/ET-SCI recommended indices, dating back to the preceding four decades, covering nearly all global land grid-cells, and assembled using a quality-controlled reanalysis data product at a high spatial resolution. Considering the computational time and resources required for assembling a comprehensive dataset of CEIs at a global scale, the biggest asset of CEI_0p25_1970_2016 from the users' perspective is the open access to a pre-compiled ready-to-use set of indices in its native data format, along with a web interface allowing robust statistical analysis and mapping of the results in a few easy steps (details in Section 5.1).

---

[12] At the time of assembling the current dataset, the newly released ECMWF-ERA5 that also includes a large set of variables was not publicly available prior to the year 2000.

4.2. Scope of Application

The CEIs included in this study are not only suited as assessment tools in multiple sectors such as Agriculture, Health, Energy, Water resource, etc., but also as metrics capable of being aggregated as composite indicators for risk assessment and vulnerability studies (e.g., as demonstrated and applied recently by [25] over Italy in the form of a "Climate Risk Index"). A number of earlier studies have demonstrated the efficacy of the CEIs, both in detection and attribution studies, as well in the impacts assessment of climate change and variability in key sectors. Examples include (i) [26] who use "Rx1day" (Table S2) to examine the changes in model-simulated extreme precipitation by decomposing the daily regional-scale extreme precipitation as contributions from atmospheric thermodynamics and dynamics; and (ii) [27] who consider a broad range of CEIs (from Tables S1 and S2), for assessing future climate change impacts on agriculture, human health, ecological ecosystems and utility (energy demand) in Canada.

Moreover, it is widely known and established in sectoral impact studies employing empirical methods that a large proportion of variation in the outcome variable is better explained by the climatic variables accounting for moderate or severe extremes (e.g., the relationship (i) between crop productivity and a variant of the index "GDDgrown" in Table S1, known as killing degree days (KDDs) [28], (ii) between electricity consumption and degree-day indices namely "CDD" and "HDD" [29]). CEI_0p25_1970_2016 for instance provides an instant resource platform for empirical modellers to download and investigate a number of potential predictor variables that are robust moderate/severe extreme indicators.

The robust characteristics and climatological attributes captured by ETCCDI/ET-SCI indices can facilitate consistent comparison of results across different climatic zones, different time periods, and the identification of regions (clusters) with similar characteristics in extremes (e.g., grid cells with similar trends in annual days when daily maximum temperature is at least 30 °C ("TXge30", Table S1). The identification of common hot spots can be of potential interest to policymakers, insurance companies, and country planners for the assessment of the risk and vulnerability of regions to extreme weather disasters (e.g., flooding, drought, heat waves).

While the mean climatology of a location is invariably well-captured by the state-of-the-art reanalysis data products and Earth System Models (ESMs), extremes (particularly in precipitation) at fine spatial scales have been difficult to replicate [30]. CEIs provide the modelling community with a detailed set of indicators enabling the comparison of different input data sources in their ability to model extremes [8,9].

Finally, with the planned inclusion of additional indices to the current inventory of ETCCDI/ET-SCI indices in the near future [9], the development of larger CEI datasets for historical and future time periods could make valuable instruments available to researchers, policymakers, and adaptation planners focusing on occurrences and return periods of rarer extreme meteorological events (e.g., using extreme value theory).

4.3. Limitations of Indices Included in CEI_0p25_1970_2016

While the CEIs included in this study (Tables S1 and S2) were developed by the WMO expert teams to largely address the growing demands of sectoral impact modellers, certain limitations of the existing ETCCDI/ET-SCI indices have been recognized, and efforts are ongoing to develop other robust indices meeting multi-sectoral requirements [9]. For instance, under the current framework of ET-SCI definitions, the Heat Wave Magnitude (HWM) indices (Table S2 are based on the methodology developed by either [31] or [32]. The more recently developed HWM Index daily (HWMId) defined by [33] and implemented in various sectoral studies (e.g., [34] for river discharge and [35] for assessing impacts on wheat yields[13]) is yet to be included in the inventory of ETCCDI/ET-SCI indices.

Moreover, the ETCCDI/ET-SCI indices are defined largely at annual timescales, and some are defined at monthly timescales as well. For certain sectoral applications (e.g., in Agriculture and Energy), the current set of monthly/annual indices may prove less useful, as climate anomalies need to be computed over different timescales. For instance, the "GSL" index (Table S1) in its current form defined at annual timescales does not account for heterogeneity in the length of crop-specific growing season (further details in [35]). In such cases, using indices computed at annual timescales can lead to misleading results. Some further shortcomings of the existing ETCCDI/ET-SCI indices are discussed and recommended for future work (details in Section 5.2).

Lastly, it must be emphasized that because CEI_0p25_1970_2016 utilizes temperature and precipitation data from GLDAS, when using the current set of indices users should keep in mind the known uncertainties and limitations of the GLDAS data (as discussed in Section 3.2).

5. Dataset Availability and Plans for Future Work

5.1. Data Access, File Naming Convention, and Size

CEI_0p25_1970_2016 can be accessed as individual netCDF4 files from https://doi.org/10.1594/PANGAEA.898014[14]. The files follow the naming convention CEI_timescale_GLDAS_0p25_deg_hist_1970_2016.nc (Figure 1), wherein "CEI" is the abbreviation of the index (as described in Tables S1 and S2) and "timescale" is either "ANN", "MON", or "DAY", relating to annual, monthly, or daily timescales[15] over which the corresponding CEI is computed.

The size of the individual NetCDF files vary between 156 megabytes (Mb) and 1.9 gigabytes (Gb), depending on the CEI and time-scales at which it is computed. One exception is the file "hw_ANN_GLDAS_0p25_deg_hist_1970_2016.nc" which is 3.1 Gb as it includes twenty individual indices in a single netCDF4 file. GLDAS does not include data over (or near) water bodies. Such grid cells where the required GLDAS TX, TN, and PR data are not available for computing the CEIs are identified by missing values "1.e+20f". Further details of the variables/dimensions in the individual netCDF4 files can be examined using either NCO or CDO commands, such as "ncdump -h netcdf_file_name" or "cdo sinfo netcdf_file_name", respectively. For creating quick plots and exploratory data analysis of individual netCDF files, open-access data tools such as Panoply (https://www.giss.nasa.gov/tools/panoply/) or NCview (http://meteora.ucsd.edu/~pierce/ncview_home_page.html) are recommended. Sample plots using Panoply for the four indices ("TXx", "HWM_Tx90", "CSDI", and "PRCPTOT") are shown in Appendix A (Figures A1–A4).

---

13   The authors use a slightly modified version of HWDId in their study, which they refer to as Heat Magnitude Day (HMD) in agriculture.

14   The dataset will also be mirrored on KNMI Climate Explorer (http://climexp.knmi.nl/about.cgi?id=someone@somewhere), a web application interface that can facilitate not only rapid aggregation and robust statistical analysis of the CEI, but also downloading of spatio-temporal subsets and quick plotting.

15   The dataset includes a total of 89 netCDF4 files (49 on annual, 39 on monthly and 1 on daily timescales). Some indices have data both on monthly and annual timescales.

File name	File format	File size [kByte]	URL file
cdd_ANN_GLDAS_0p25_deg_hist_1970_2016.nc	NC	158721	Link
cdd_MON_GLDAS_0p25_deg_hist_1970_2016.nc	NC	1905590	Link
csdi5_ANN_GLDAS_0p25_deg_hist_1970_2016.nc	NC	158721	Link
csdi_ANN_GLDAS_0p25_deg_hist_1970_2016.nc	NC	158721	Link
cwd_ANN_GLDAS_0p25_deg_hist_1970_2016.nc	NC	158721	Link
cwd_MON_GLDAS_0p25_deg_hist_1970_2016.nc	NC	1905590	Link
dtr_ANN_GLDAS_0p25_deg_hist_1970_2016.nc	NC	158721	Link
dtr_MON_GLDAS_0p25_deg_hist_1970_2016.nc	NC	1905590	Link
fd_ANN_GLDAS_0p25_deg_hist_1970_2016.nc	NC	158717	Link
fd_MON_GLDAS_0p25_deg_hist_1970_2016.nc	NC	1905586	Link
gddgrow10_ANN_GLDAS_0p25_deg_hist_1970_2016.nc	NC	158721	Link
gsl_ANN_GLDAS_0p25_deg_hist_1970_2016.nc	NC	158721	Link
hw_ANN_GLDAS_0p25_deg_hist_1970_2016.nc	NC	3173596	Link
id_ANN_GLDAS_0p25_deg_hist_1970_2016.nc	NC	158717	Link
id_MON_GLDAS_0p25_deg_hist_1970_2016.nc	NC	1905586	Link
prcptot_ANN_GLDAS_0p25_deg_hist_1970_2016.nc	NC	158721	Link
prcptot_MON_GLDAS_0p25_deg_hist_1970_2016.nc	NC	1905590	Link
r10mm_ANN_GLDAS_0p25_deg_hist_1970_2016.nc	NC	158721	Link
r10mm_MON_GLDAS_0p25_deg_hist_1970_2016.nc	NC	1905590	Link

Figure 1. Snapshot of the data repository web interface showing individual files in CEI_0p25_1970_2016 that are available for download at https://doi.org/10.1594/PANGAEA.898014.

5.2. Ongoing Work and Recommendations for Work in Future

The indices in CEI_0p25_1970_2016 are intended to be updated post-2016 years, subject to the availability of the required GLDAS raw meteorological variables in the coming years. The updated longer time-series of CEIs of more recent years should prove beneficial to the research community focusing on recent extreme events (e.g., the droughts of 2017 and 2018 in south-east Australia, the heat waves of 2018 in California, United States of America, the more recent January–February 2019 extreme cold wave in North America). Additionally, upon the formal inclusion of any new indices (such as the "HWMId" and the "Crop-specific GSL" as discussed in Section 4.3) by the WMO expert teams to their list of ET-SCI indices, the same will be formally included in the existing dataset presented in this study.

While the ETCCDI/ET-SCI core and non-core indices employed in this study encompass a very large spectrum of sectoral and non-sectoral indices, the list is by no means exhaustive. Motivated by the suggestions of the R ClimPACT2 [9] package creators, another dataset of indices largely relevant for health and energy sectors (called "HEI_0p25_1970_2016") is currently under preparation [12].

Some features of HEI_0p25_1970_2016 will for instance be the inclusion of the two ETCCDI indices (i.e., CDD and HDD [36]) that are not included in this study[16]. Moreover, HEI_0p25_1970_2016 will also account for additional meteorological variables (e.g., near-surface relative humidity and wind speed) for computing non ETCCDI/ET-SCI indices, such as the Humidex [37,38], the Heat Index (HI) [39,40], and the Discomfort Index (DI) [41,42]. Together, both CEI_0p25_1970_2016 and HEI_0p25_1970_2016 are aimed to address the growing needs of the climate impact community, by overcoming the current data scarcity of high-resolution global gridded CEIs in earth science.

---

[16] The R ClimPACT2 used in the present study for computing CEI_0p25_1970_2016 is hard-coded to compute the degree-days (CDD, HDD) on annual time scales. Degree-days at monthly and seasonal timescales are equally important in the energy sector. These are developed at various base (threshold) temperatures at the same gridded resolution in HEI_0p25_1970_2016.

Supplementary Materials: The following are available online at http://www.mdpi.com/2306-5729/4/1/41/s1, Table S1: 32 Core ET-SCI indices. Bold indicates index is also an ETCCDI index. (TX: daily maximum near-surface air temperature, TN: daily minimum near-surface air temperature, PR: daily near-surface total precipitation, H: Health, AFS: Agriculture and Food Security, WRH: Water Resources and Hydrology); Table S2: 39 non-core ET-SCI indices. Bold indicates index is also an ETCCDI index. Sectoral abbreviations same as in Table S1.

Funding: This work was supported by research funding from the European Research Council (ERC) under the European Union's Horizon 2020 research and innovation programme under grant agreement No. 756194 (ENERGYA).

Acknowledgments: The author is grateful to Nicholas Herold in assisting with the R software package ClimPACT2; Lisa Alexander and Enrico Scoccimarro for constructive discussion on sectoral extreme indices; Enrica De Cian for feedback on the draft version of the paper; the high-performance computing resources of the Boston University Shared Computing Cluster (SCC) on which the CEIs were computed; and NASA Goddard Earth Sciences Data and Information Services Center (GES DISC) for making GLDAS data publicly available. Developers of R SPEI package, CDO, and NCO are also acknowledged for providing open-access tools that were used for data preparation in this study. The constructive feedback received from three anonymous reviewers helped to improve the manuscript further.

Conflicts of Interest: The author declares no conflict of interest.

Appendix A. Sample Plots of Selective Indices from Tables S1 and S2 Using Panoply

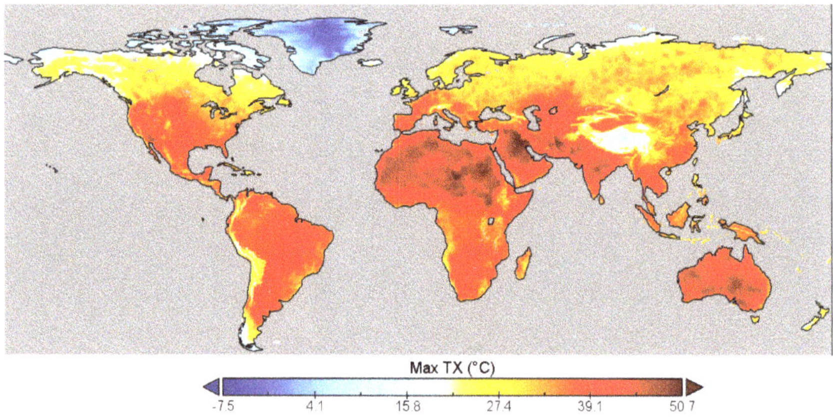

Figure A1. Annual Warmest Day "TXx" (°C) in 2003.

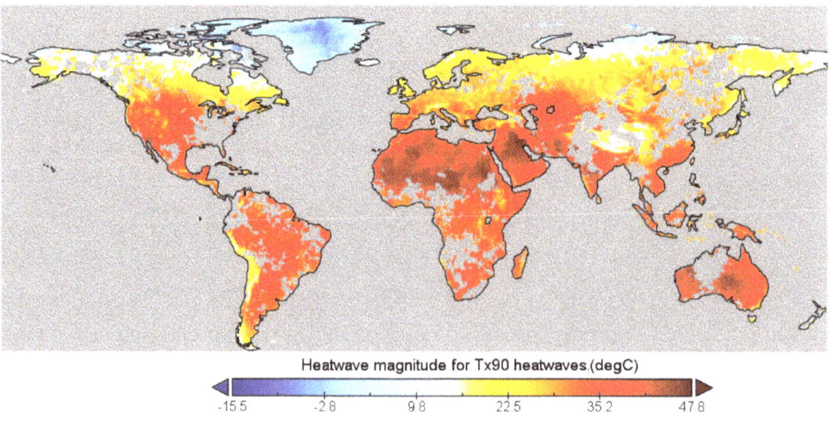

Figure A2. Warmest Day "HWM_Tx90" (°C) in 2003 (Average temperature across all individual heatwaves).

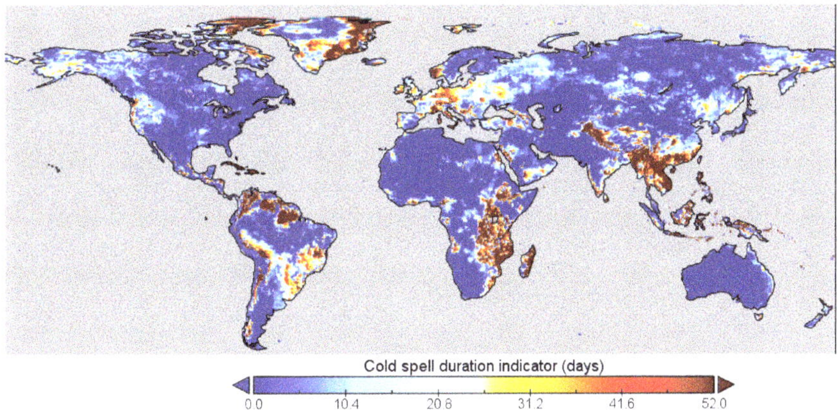

Figure A3. Cold Spell Duration Index "CSDI" (Days) in 2013.

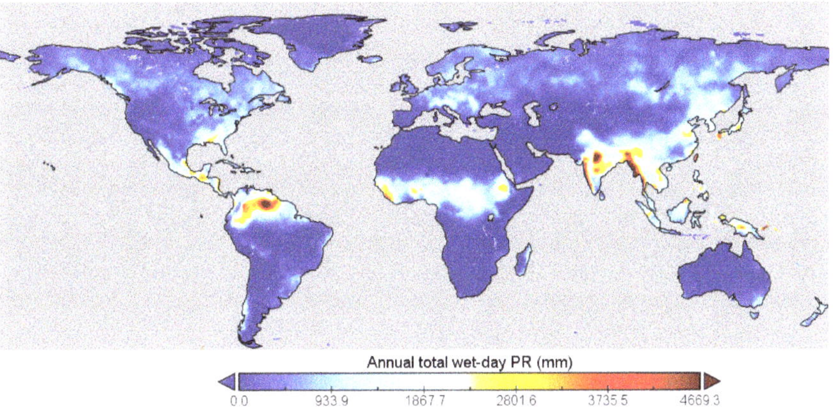

Figure A4. Total wet-day rainfall "PRCPTOT" (mm) in July 2005.

References

1. Easterling, D.R.; Meehl, G.A.; Parmesan, C.; Changnon, S.A.; Karl, T.R.; Mearns, L.O. Climate Extremes: Observations, Modeling, and Impacts. *Science* **2000**, *289*, 2068–2074. [CrossRef] [PubMed]
2. Alexander, L.V. Global observed long-term changes in temperature and precipitation extremes: A review of progress and limitations in IPCC assessments and beyond. *Weather Clim. Extrem.* **2016**, *11*, 4–16. [CrossRef]
3. Dosio, A. Projections of climate change indices of temperature and precipitation from an ensemble of bias-adjusted high-resolution EURO-CORDEX regional climate models. *J. Geophys. Res. Atmos.* **2016**, *121*, 5488–5511. [CrossRef]
4. Karl, T.R.; Nicholls, N.; Ghazi, A., CLIVAR/GCOS/WMO Workshop on Indices and Indicators for Climate Extremes Workshop Summary. In *Weather and Climate Extremes: Changes, Variations and a Perspective from the Insurance Industry*; Karl, T.R., Nicholls, N., Ghazi, A., Eds.; Springer: Dordrecht, The Netherlands, 1999; pp. 3–7. [CrossRef]
5. Alexander, L.; Tebaldi, C. Chapter 10 - Climate and Weather Extremes: Observations, Modelling, and Projections. In *The Future of the World's Climate*, 2nd ed.; Henderson-Sellers, A., McGuffie, K., Eds.; Elsevier: Boston, MA, USA, 2012; pp. 253–288.
6. Donat, M.G.; Alexander, L.V.; Yang, H.; Durre, I.; Vose, R.; Dunn, R.J.H.; Willett, K.M.; Aguilar, E.; Brunet, M.; Caesar, J.; et al. Updated analyses of temperature and precipitation extreme indices since the beginning of the twentieth century: The HadEX2 dataset. *J. Geophys. Res. Atmos.* **2013**, *118*, 2098–2118. [CrossRef]

7. Donat, M.G.; Alexander, L.V.; Yang, H.; Durre, I.; Vose, R.; Caesar, J. Global Land-Based Datasets for Monitoring Climatic Extremes. *Bull. Am. Meteorol. Soc.* **2013**, *94*, 997–1006. [CrossRef]

8. Sillmann, J.; Kharin, V.V.; Zhang, X.; Zwiers, F.W.; Bronaugh, D. Climate extremes indices in the CMIP5 multimodel ensemble: Part 1. Model evaluation in the present climate. *J. Geophys. Res. Atmos.* **2013**, *118*, 1716–1733. [CrossRef]

9. Alexander, L.; Herold, N. ClimPACT2 Indices and Software (R Software Package). Available online: https://htmlpreview.github.io/?https://raw.githubusercontent.com/ARCCSS-extremes/climpact2/master/user_guide/ClimPACT2_user_guide.htm (accessed on 12 March 2019).

10. Alexander, L.V.; Zhang, X.; Peterson, T.C.; Caesar, J.; Gleason, B.; Klein Tank, A.M.G.; Haylock, M.; Collins, D.; Trewin, B.; Rahimzadeh, F.; et al. Global observed changes in daily climate extremes of temperature and precipitation. *J. Geophys. Res. Atmos.* **2006**, *111*. [CrossRef]

11. Zhang, X.; Alexander, L.; Hegerl, G.C.; Jones, P.; Tank, A.K.; Peterson, T.C.; Trewin, B.; Zwiers, F.W. Indices for monitoring changes in extremes based on daily temperature and precipitation data. *Wiley Interdiscip. Rev. Clim. Chang.* **2011**, *2*, 851–870. [CrossRef]

12. Mistry, M. A High-Resolution Global Gridded Dataset of Climate Indices Relevant for Health and Energy Sector. **2019**, under prep.

13. Rodell, M.; Houser, P.R.; Jambor, U.; Gottschalck, J.; Mitchell, K.; Meng, C.J.; Arsenault, K.; Cosgrove, B.; Radakovich, J.; Bosilovich, M.; et al. The Global Land Data Assimilation System. *Bull. Am. Meteorol. Soc.* **2004**, *85*, 381–394. [CrossRef]

14. Ji, L.; Senay, G.B.; Verdin, J.P. Evaluation of the Global Land Data Assimilation System (GLDAS) Air Temperature Data Products. *J. Hydrometeorol.* **2015**, *16*, 2463–2480. [CrossRef]

15. Iizumi, T.; Takikawa, H.; Hirabayashi, Y.; Hanasaki, N.; Nishimori, M. Contributions of different bias-correction methods and reference meteorological forcing data sets to uncertainty in projected temperature and precipitation extremes. *J. Geophys. Res. Atmos.* **2017**, *122*, 7800–7819. [CrossRef]

16. R Core Team. *R: A Language and Environment for Statistical Computing*; R Core Team: Vienna, Austria, 2018.

17. Begueria, S.; Vicente-Serrano, S.M. SPEI: Calculation of the Standardised Precipitation-Evapotranspiration Index. R Package Version 1.6. 2013. Available online: http://CRAN.R-project.org/package=SPEI (accessed on 11 August 2018).

18. Kumar, S.V.; Peters-Lidard, C.D.; Tian, Y.; Houser, P.R.; Geiger, J.; Olden, S.; Lighty, L.; Eastman, J.L.; Doty, B.; Dirmeyer, P.; et al. Land information system: An interoperable framework for high resolution land surface modeling. *Environ. Model. Softw.* **2006**, *21*, 1402–1415. [CrossRef]

19. Peters-Lidard, C.D.; Houser, P.R.; Tian, Y.; Kumar, S.V.; Geiger, J.; Olden, S.; Lighty, L.; Doty, B.; Dirmeyer, P.; Adams, J.; Mitchell, K.; Wood, E.F.; Sheffield, J. High-performance Earth system modeling with NASA/GSFC's Land Information System. *Innov. Syst. Softw. Eng.* **2007**, *3*, 157–165. [CrossRef]

20. Zender, C.S. Analysis of self-describing gridded geoscience data with netCDF Operators (NCO). *Environ. Model. Softw.* **2008**, *23*, 1338–1342. [CrossRef]

21. Schulzweida, U. *Climate Data Operators (CDO) User Guide, Version 1.9.0*; Max-Planck-Institute for Meteorology: Hamburg, Germany, 2018.

22. De Cian, E.; Sue Wing, I. Global Energy Consumption in a Warming Climate. *Environ. Resour. Econ.* **2019**, *72*, 365–410. [CrossRef]

23. Zhong, L.; Su, Z.; Ma, Y.; Salama, M.S.; Sobrino, J.A. Accelerated Changes of Environmental Conditions on the Tibetan Plateau Caused by Climate Change. *J. Clim.* **2011**, *24*, 6540–6550. [CrossRef]

24. Gao, Y.; Cuo, L.; Zhang, Y. Changes in Moisture Flux over the Tibetan Plateau during 1979–2011 and Possible Mechanisms. *J. Clim.* **2014**, *27*, 1876–1893. [CrossRef]

25. Mysiak, J.; Torresan, S.; Bosello, F.; Mistry, M.; Amadio, M.; Marzi, S.; Furlan, E.; Sperotto, A. Climate risk index for Italy. *Philos. Trans. Ser. A Math. Phys. Eng. Sci.* **2018**, *376*, 20170305. [CrossRef] [PubMed]

26. Pfahl, S.; O'Gorman, P.A.; Fischer, E.M. Understanding the regional pattern of projected future changes in extreme precipitation. *Nat. Clim. Chang.* **2017**, *7*, 423. [CrossRef]

27. Li, G.; Zhang, X.; Cannon, A.J.; Murdock, T.; Sobie, S.; Zwiers, F.; Anderson, K.; Qian, B. Indices of Canada's future climate for general and agricultural adaptation applications. *Clim. Chang.* **2018**, *148*, 249–263. [CrossRef]

28. Schlenker, W.; Roberts, M.J. Nonlinear temperature effects indicate severe damages to U.S. crop yields under climate change. *Proc. Natl. Acad. Sci. USA* **2009**, *106*, 15594–15598. [CrossRef]

29. Guan, H.; Beecham, S.; Xu, H.; Ingleton, G. Incorporating residual temperature and specific humidity in predicting weather-dependent warm-season electricity consumption. *Environ. Res. Lett.* **2017**, *12*, 024021. [CrossRef]

30. Sillmann, J.; Thorarinsdottir, T.; Keenlyside, N.; Schaller, N.; Alexander, L.V.; Hegerl, G.; Seneviratne, S.I.; Vautard, R.; Zhang, X.; Zwiers, F.W. Understanding, modeling and predicting weather and climate extremes: Challenges and opportunities. *Weather Clim. Extrem.* **2017**, *18*, 65–74. [CrossRef]

31. Perkins, S.E.; Alexander, L.V. On the Measurement of Heat Waves. *J. Clim.* **2013**, *26*, 4500–4517. [CrossRef]

32. Nairn, J.R.; Fawcett, R.J.B. The Excess Heat Factor: A Metric for Heatwave Intensity and Its Use in Classifying Heatwave Severity. *Int. J. Environ. Res. Public Health* **2015**, *12*, 227–253. [CrossRef] [PubMed]

33. Russo, S.; Sillmann, J.; Fischer, E.M. Top ten European heatwaves since 1950 and their occurrence in the coming decades. *Environ. Res. Lett.* **2015**, *10*, 124003. [CrossRef]

34. Zampieri, M.; Russo, S.; di Sabatino, S.; Michetti, M.; Scoccimarro, E.; Gualdi, S. Global assessment of heat wave magnitudes from 1901 to 2010 and implications for the river discharge of the Alps. *Sci. Total Environ.* **2016**, *571*, 1330–1339. [CrossRef] [PubMed]

35. Zampieri, M.; Ceglar, A.; Dentener, F.; Toreti, A. Wheat yield loss attributable to heat waves, drought and water excess at the global, national and subnational scales. *Environ. Res. Lett.* **2017**, *12*, 064008. [CrossRef]

36. *ASHRAE Handbook*; American Society of Heating, Refrigerating and Air-Conditioning Engineers: Atlanta, GA, USA, 2001; Chapter 31.

37. Masterton, J.M.; De l'environnement atmosphérique, C.S.; Richardson, F.A. *Humidex: A Method of Quantifying Human Discomfort Due to Excessive Heat and Humidity*; Environment Canada, Atmospheric Environment: Downsview, ON, Canada, 1979.

38. Buzan, J.R.; Oleson, K.; Huber, M. Implementation and comparison of a suite of heat stress metrics within the Community Land Model version 4.5. *Geosci. Model Dev.* **2015**, *8*, 151–170. [CrossRef]

39. Steadman, R.G. The Assessment of Sultriness. Part I: A Temperature-Humidity Index Based on Human Physiology and Clothing Science. *J. Appl. Meteorol.* **1979**, *18*, 861–873. [CrossRef]

40. Rothfusz, L. The Heat Index "Equation" (or, More Than You Ever Wanted to Know About Heat Index); *Natl. Weather. Serv. Tech. Attach.* **1990**. Available online: https://www.weather.gov/media/bgm/ta_htindx.PDF (accessed on 12 March 2019).

41. Thom, E.C. The Discomfort Index. *Weatherwise* **1959**, *12*, 57–61. [CrossRef]

42. Epstein, Y.; Moran, D.S. Thermal Comfort and the Heat Stress Indices. *Ind. Health* **2006**, *44*, 388–398. [CrossRef] [PubMed]

Review

A Lack of "Environmental Earth Data" at the Microhabitat Scale Impacts Efforts to Control Invasive Arthropods That Vector Pathogens

Emily L. Pascoe [1],*, Sajid Pareeth [2], Duccio Rocchini [3,4,5,6] and Matteo Marcantonio [7],*

[1] Department of Medicine and Epidemiology, School of Veterinary Medicine, University of California, Davis, CA 95616, USA

[2] Water Science and Engineering Department, IHE Delft Institute for Water Education, 2611AX Delft, The Netherlands; s.pareeth@un-ihe.org

[3] Center Agriculture Food Environment (C3A), University of Trento, Via E. Mach 1, 38010 S. Michele all'Adige, TN, Italy; duccio.rocchini@unitn.it

[4] Department of Cellular, Computational and Integrative Biology (CIBIO), University of Trento, Via Sommarive, 14, 38123 Povo, TN, Italy

[5] Department of Biodiversity and Molecular Ecology, Fondazione Edmund Mach, Research and Innovation Centre, Via E. Mach 1, 38010 S. Michele all'Adige, TN, Italy

[6] Department of Applied Geoinformatics and Spatial Planning, Faculty of Environmental Sciences, Czech University of Life Sciences Prague, Kamýcka 129, Praha—Suchdol 16500, Czech Republic

[7] Department of Pathology, Microbiology, and Immunology, School of Veterinary Medicine, University of California, Davis, CA 95616, USA

* Correspondence: elpascoe@ucdavis.edu (E.L.P.); matmarcantonio@ucdavis.edu (M.M.)

Received: 27 May 2019; Accepted: 19 September 2019; Published: 29 September 2019

Abstract: We currently live in an era of major global change that has led to the introduction and range expansion of numerous invasive species worldwide. In addition to the ecological and economic consequences associated with most invasive species, invasive arthropods that vector pathogens (IAVPs) to humans and animals pose substantial health risks. Species distribution models that are informed using environmental Earth data are frequently employed to predict the distribution of invasive species, and to advise targeted mitigation strategies. However, there are currently substantial mismatches in the temporal and spatial resolution of these data and the environmental contexts which affect IAVPs. Consequently, targeted actions to control invasive species or to prepare the population for possible disease outbreaks may lack efficacy. Here, we identify and discuss how the currently available environmental Earth data are lacking with respect to their applications in species distribution modeling, particularly when predicting the potential distribution of IAVPs at meaningful space-time scales. For example, we examine the issues related to interpolation of weather station data and the lack of microclimatic data relevant to the environment experienced by IAVPs. In addition, we suggest how these data gaps can be filled, including through the possible development of a dedicated open access database, where data from both remotely- and proximally-sensed sources can be stored, shared, and accessed.

Keywords: arthropod vector; invasive species; microhabitat; species distribution modeling; remote sensing

---

1. Introduction

In an era of major global change (i.e., in climate, land use, habitat fragmentation, and movements of humans and other species) the introduction of invasive species and the geographic expansion of endemic species to novel ranges are occurring at unprecedented rates [1,2]. Invasive species have extensive negative impacts on the ecosystems they invade, such as losses in both taxonomic and functional diversity [3], resulting in severe economic consequences. For example, in the USA invasive insects cost the agricultural sector USD 13 billion per year due to crop loss and damage [4], routine activities to control *Aedes* mosquitoes in Cuba cost USD 16.80 per household [5], and Great Britain spends USD 34.6 million per year on the control of invasive fresh-water species [6]. The stakes are even higher when invasive species can vector pathogens that cause disease in humans, animals, or plants. Some arthropod species that are highly invasive are among the most effective vectors of human pathogens. Mosquitoes, such as *Aedes aegypti* [Linneaus, 1762] and *Aedes albopictus* [Skuse, 1895], are invasive arthropods able to vector pathogens (henceforth referred to as IAVPs) and transmit three globally important viruses to humans: chikungunya, dengue, and Zika [7]. Likewise, the Asian longhorned tick *Haemaphysalis longicornis* [Neumann, 1901], which is rapidly expanding across the east coast of the USA [8], can vector the severe fever and thrombocytopenia syndrome virus, which has human fatality rates exceeding 30% in Asia [9]. Of veterinary importance, the biting midge *Culicoides imicola* [Kieffer, 1913] is currently expanding in range throughout Europe and can transmit the bluetongue and African horse sickness viruses [10]. Plants are also affected, cotton whiteflies (*Bemisia* species, including *Bemisia tabaci* [Gennadius, 1889]), now present in every continent except Antarctica, can transmit over 100 different plant viruses [10].

In order to prevent the potentially catastrophic ecological, economical, and health consequences associated with IAVPs, mitigation methods must be rapidly employed following species introduction or expansion into a new geographical range [11]. Mitigation methods may include IAVP control and eradication, or communication of the risks to policy makers, physicians and the public, and environmental data are often used to inform these different processes. Here, we describe the benefits and limitations associated with using i) remotely sensed data, which we define as data acquired by sensors mounted on satellite, airborne, or other distant means, and ii) proximally sensed data, which we define as having been collected by a ground-based, or other platform, in close proximity to the variable being measured, in order to inform IAVP mitigation.

2. Linking Environmental Earth Data and IAVPs

As observed by Malanson and Walsh; "*detection and eradication [of invasive species] are essentially spatial problems. They primarily require learning where the invasives are and getting there*" [12]. This is a simplification of a more complex issue, which may also involve a lack of personnel or funding to efficiently implement detection and eradication, insufficient communication or perception of the IAVP risk, and even IAVP resistance to control measures. However, environmental data can be used to address the "spatial problems" by informing predictions on where invasive species may be introduced and become established.

In some instances, using environmental data in the mitigation of an invasive species can be as straightforward as directly detecting the species. For example, thanks to the reflectance properties of vegetation, invasive plants can be mapped using indices such as NDVI (the Normalized Difference Vegetation Index) or EVI (the Enhanced Vegetation Index) that are derived from remotely sensed data that measures infrared reflectance (e.g., in [13], and also see [14] for a review on this method). A similar concept can be applied to invasive arthropods which cause damage to vegetation, and NDVI data has been used to track the dispersal of invasive insects by monitoring defoliation [15,16]. Although weather radars have detected mass migrations of invasive insects [17], as yet remotely sensed data cannot directly characterize IAVP geographical distributions. There are promising proximal sensing methods that use reflectance data from cameras that can detect and differentiate between multiple fruit fly species, including those that vector crop pathogens [18] (see also [19] for an interesting application of proximal sensing of an invasive pathogenic plant bacterium). However, mapping IAVP distribution in real-time is often less desirable than preempting the potential geographic distribution, as surveillance and control are more efficient if implemented prior to the establishment of a species [11,12,20].

Species distribution models (SDMs) are frequently used to predict the current and future geographic distribution of IAVPs [11,12,20] due to their ability to be applied to species that cannot be directly detected because they are small, elusive or inhabit remote locations. Typically, correlative SDMs apply an algorithm, such as maximum entropy, boosted regression trees or random forest, that combines empirical occurrence data on the species with relevant environmental data (e.g., average temperature and precipitation) to predict the spatial and temporal distributions of a species [21,22]. Mechanistic models, such as compartmental or agent-based models have also been developed, alone or in combination with correlative models, to characterize potential species distributions [23,24]. Here, we adopt a broad definition of SDMs to include any modeling approach that aims to predict the distribution of a species, from logistic regression to multi-criteria decision analyses. In the last few decades there has been a sharp increase in the number of publications on SDMs, with hundreds published each year [25]. This dramatic rise in interest in SDMs is in part due to advances in remote sensing technology, including new satellites and sensors that have hugely increased the quantity and quality of environmental data that can be used [26].

3. Gathering Environmental Data for SDMs

The accuracy of SDM predictions is highly dependent on how closely the data used in the model match conditions relevant to the species and, despite considerable increases in both spatial and temporal resolution of available environmental data, there is often still a substantial mismatch in the conditions represented by the available data and those experienced by IAVP species. Environmental data used in SDMs can be classified as bio-physical or climatic, both of which can be measured by proximal sensing, but data used in SDMs is typically derived from remote sensing.

3.1. Bio-Physical Variables

Bio-physical variables generally include land-use, land cover, primary productivity, and vegetation phenology and fragmentation. Bio-physical variables are almost exclusively derived from Earth observation satellites which measure either reflectance at various wavelengths in the electromagnetic spectrum, or emitted radiances in the thermal spectrum. These reflectance data can be used to calculate NDVI and NDWI (Normalized Difference Water Index), which are applied instead of, or alongside, other satellite imagery/reflectance data to ascertain variables such as land-use and land cover (Table 1). Satellite data are available in a wide range of spatial (<1 m to >5 km) and temporal (hourly to yearly) resolutions, and allow for some user flexibility based on the scale at which the model is applied (e.g., eco-region, county, national, global). Given technological limitations due to on-board storage media or limited opportunity for data transmission, spatial and temporal resolution of remote sensing tools are inversely correlated [27]. As the majority of bio-physical variables remain static or exhibit very gradual changes over time, spatial resolution is often prioritized over temporal resolution. For example, since 1972 the NASA-USGS Landsat series has provided uninterrupted data on the Earth's surface at a relatively high resolution of 30 m, but measurements are only taken once every 16 days, although this will increase to every eight days starting from 2020. NOAA VIIRS provides a series of environmental data, as well as monthly cloud-free composites of visible infrared emittance for the entire Earth during night at a resolution of 15 arcsec (<500 m at the equator) [28], which can be used as a proxy for human settlements to inform the possible human contact risk associated with IAVP presence [29,30]. Since the 1980's, satellite remote sensors such as AVHRR and, many years later, MODIS, have allowed the derivation of more spatially and temporally continuous vegetation and surface temperature data at a moderate spatial resolution (250–1000 m), but with more frequent (daily) observations, thus greatly enriching the available datasets [31]. In addition, the more recent Sentinel missions (2A, 2B, 3) from the European Space Agency (ESA) have offered optical data at 10–300 m spatial resolution every 3–7 days since 2016.

Table 1. A summary of the main sources of environmental data that can be used in species distribution modeling of invasive species that vector pathogens, including information on the environmental variables that can be derived from the data, spatial and temporal resolution, and the geographic extent that the data are able to cover.

Mission/Sensor	Type of Sensing	Environmental Variables for IAVPs	Spatial Resolution	Temporal Grain and Extent	Extent
NASA-MODIS	Multi-satellite	NDVI, NDWI, LST, Land cover	0.25–1 km	4-times/day [2001–present]	Global
NASA-USGS Landsat series	Multi-satellite	NDVI, NDWI, imagery	30 m	16 d [1972–present]	Global
ESA SENTINEL missions	Multi-satellite	NDVI, NDWI, LST, imagery	10–300 m	3–10 d [2015–present]	Global
NOAA VIIRS	Multi-satellite	NDVI, NDWI, LST, imagery, human settlements	375–750 m	1 d–monthly [2015–present]	Global
Global Precipitation Measurement Mission (GPMM)	Multi-satellite	Precipitation	11 km*	2–3 h [2015–present]	Global (65S-65N)
Tropical Rainfall Measuring Mission (TRMM)	Multi-satellite	Rainfall	28 km*	3 h–7 d [1998–2015]	Tropical and sub-tropical regions
USDA-NAIP	Airborne	NDVI, imagery	60 cm– 1 m	"Snapshot" every 3 years [2003–present]	Mainland USA (variable coverage)
Dataset Name	**Ancillary Data**	**Environmental Variables for IAVPs**	**Spatial Resolution**	**Temporal Grain and Extent**	**Extent**
WordClim	Weather station	2 m air temperature and precipitation	1 km*	LTA 1950–2000	Global
MODIS Land Cover Type/Dynamics	Satellite	Land cover	0.5–1 km	Yearly/twice a year [2001–present)	Global
Copernicus Land Cover	Multi-satellite (SPOT, PROBA-V, Sentinel-2)	Land cover	100 m	Multi-year [2015–present)	Global
USGS Land Cover maps	Satellite (Landsat) and geospatial ancillary datasets	Land cover/impervious surface	30 m	Multi-year [2001–present]	Continental US
CORINE Land Cover maps	Multi-satellite (Landsat, SPOT, IRS, RapidEye, Sentinel-2)	Land cover	100 m	Multi-year [1990–present]	Extended EU
PRISM Climate data	Weather station	Air temperature, precipitation, vapor pressure, day length	0.8–4 km	Daily [1895–**)	Continental US
Daily Surface weather and climatological summaries (DAYMET)	Weather station	Air temperature, precipitation, vapor pressure, day length	1 km	Daily [1980–present calendar year]	North America, Puerto Rico and Hawaii

* Approximate spatial resolution in metric unit of lengths on the equator, ** 6 months from current date.

The accuracy of satellite data is generally strongly linked to the method of derivation, geographical region, climatic condition, and availability of in-situ data for calibration, which in turn affect SDM results. For example, cloud cover often hinders satellite optical data, especially in inter-tropical regions, but there are multiple statistical approaches that can fill these gaps over space or time [32]. Remotely sensed data on bio-physical Earth observations can be combined with ground-based (in-situ) data to provide crucial information on habitat structure, and are therefore commonly used in SDMs. Large extent datasets for bio-physical variables, such as the Global Copernicus Land Cover maps (spatial resolution = 100 m) [33], the pan-European Corine Land Cover (100 m) [34], the USA National Land Cover Datasets (30 m) [35] and the global MODIS Land Cover Type/Dynamics (500 m–1 km) [36], are derived using a combination of satellite and ground (in-situ) sensors [35]. However, these data are typically presented as a single multi-annual "snapshot" using a composite of several observations over time, and thus, provide very limited information on temporal variation.

3.2. Climatic Variables

Climatic data, which is often fundamental in the physiology of arthropods, includes variables such as land surface temperature (LST) or air temperature and precipitation. Such data are commonly derived from remote sensing and are frequently used in SDMs. Precipitation can be measured by active satellite sensors in the micro-wave region and offer high temporal (hourly) but coarse spatial resolution data (e.g., GPM and TRMM; Table 1). As for bio-physical variables, the spatial and temporal resolution of satellite data for climatic variables are also inversely related, which results in a lack of high spatial resolution data at higher temporal frequencies of measurement. In the case of climatic variables, which can vary minute-by-minute, temporal resolution is highly important. This trade-off often plays a significant role in attaining high accuracy results from SDMs. To fill these temporal gaps, recent satellite missions that measure radiance in the thermal spectrum bands (i.e., which measure temperature) are focused on providing higher spatial resolution climatic data with frequent measurements (e.g., Sentinel 2A/B data at 10 m with weekly acquisitions).

In addition to satellite-derived Earth data, data collected by ground-based weather stations, or a combination of both, such as the WorldClim, PRISM, Daymet and ECA&D datasets, are perhaps the most widely used climatic data in SDMs due to the user friendly format that requires comparatively little pre-processing compared with satellite data (e.g., [37–42]). As weather stations measure variables at discrete geographic locations these data must be interpolated to create a continuous spatial layer before being used in SDMs. There are multiple methods by which weather station data can be interpolated, but all are limited by the density of weather stations in the study area, and are confounded by topographical features and spatial gradients, although satellite or other remotely sensed data can help to remedy some of these shortcomings [37,43,44].

For regional SDM applications high resolution datasets are required, but the availability of such data remains a challenge also for current satellite missions, despite considerable improvements during the last few years with the advent of the new Landsat and Sentinel missions.

4. Issues Faced When Using Environmental Data in IAVP Models

As described, there are many environmental datasets available that can inform SDMs. However, these datasets are often of limited relevance in the context of IAVP modeling, not least due to substantial mismatches between the spatial resolution at which predictions are made and the resolution at which the predictions are interpreted, communicated or applied. The spatial resolution of model predictions are constrained by the resolution of the environmental data used, which is typically in the order of kilometers. However, the subsequent predictions are often used to inform actions applied at spatial scales in the order of meters, such as informing which neighborhoods should be targeted for surveillance and control, where to install deer fences to control tick abundances, or communicating IAVP presence. Although some inaccuracies in SDM outputs may seem trivial in the context of a scientific paper, they can pose a serious issue when accurate predictions are required for use in "real world" scenarios. For example, SDMs and model-derived data are used by the Centers for Disease Control and Prevention (CDC) in the USA to inform administrative regions on the likelihood of *Aedes* mosquito invasion, in order to distribute vector control resources (e.g., intensive surveillance and insecticide application) [45,46]. Consequently, disparities between the spatial resolution of the data used to inform the model and that at which model outputs are applied will result in model outputs that are inaccurate for their intended applications. At best, IAVP distributions may be over-estimated, leading to unnecessary use of resources, and at worst, distributions can be underestimated such that no, or insufficient, actions are employed to control IAVPs in an area that is actually at risk. Indeed, an economic evaluation of biological invasions states that *"uncertainty prevails concerning what ecosystems will be invaded and what impacts an invasion will have within these ecosystems"*, highlighting that accurate ecological and economic analyses are crucial in the allocation of finite resources to control invasive species [47].

There is also a mismatch in spatial resolution between environmental data used in SDMs, and that at which the IAVP is affected. Indeed, many arthropods, such as mosquitoes and ticks are small and poikilothermic, and are therefore heavily affected by microclimatic conditions, which vary at fine spatial scales (in the order of centimeters to meters) and differ to the surrounding macroclimatic conditions [48–50]. For example, potentially invasive ticks, especially nidicolous (nest-dwelling) species, spend almost their entire life-cycle within a limited spatial radius; following a bloodmeal they detach from the host and remain within the host's nest or a nearby sheltered area, such as a cave or crevice, in order to metamorphose [51]. Within these isolated and sheltered microhabitats environmental conditions can be very different to those in the surrounding environment. In the same way, IAVPs can be sensitive to extreme environmental conditions, for example the lone star tick (*Amblyomma americanum* [Linnaeus, 1758]), which is invasive across much of the north east of the USA, dies within just 2 h of exposure to temperatures of ≤-3 °C in the laboratory [52] and rapidly desiccates when exposed for several hours to temperatures exceeding 30 °C [53]. Likewise, mortality of *Culicoides brevitarsis* [Kieffer, 1917] (Diptera: Ceratopogonidae), a vector of the bluetongue virus, is high in the laboratory when temperatures are greater than 35 °C, even if just for a few days [54]. Consequently, high temporal resolution of data is required to accurately capture the variance and range in environmental variables [44], but at present the most accessible remotely sensed data are only available for 1–6 day interval measurements, thus do not capture data at the same hourly temporal resolution that can affect IAVP survival. There is a wealth of literature demonstrating that if species were theoretically subjected to the macroclimate as measured by remote sensing, rather than the microclimate which they truly experience, their behavior, reproduction, growth, survival, and both phenotypic and genotypic adaptations would all be profoundly impacted [55].

In addition to issues of resolution in environmental data, some factors that impact IAVP distribution cannot be directly measured, and instead other measurements are used as a proxy, or are interpolated, for the variable of interest. Due to its importance in the IAVP life cycle, temperature is among the most broadly applied variables in IAVP species distribution modeling. However, land surface temperature is generally used as a proxy for ambient temperature [27,56,57], whilst relative humidity, which is vital to arthropod survival, is often calculated from temperature and dew point measurements, or minimum day-time air temperature [58]. SDMs are made further complex when the species of interest has multiple life stages, each of which may exploit a different microhabitat. Mosquitoes have an "amphibious" life history, throughout which they experience air, below-water, and water-surface temperatures, by having terrestrially fixed or floating eggs, aquatic immature larvae and flying adults [59]. Researchers have measured air temperature, water temperature, and precipitation to understand whether air temperature, usually used to determine mosquito distribution or life cycle, provides an appropriate direct measure for determining *Anopheles* [Diptera: Culicidae] larval development in water [48]. The authors of one such study concluded that their results *"suggest that although widely used, air temperature alone does not provide an appropriate variable for estimating immature mosquito development or for setting threshold temperatures"*. Another study that measured temperature in microhabitats suitable for *Aedes* mosquitoes found that when utilizing temperature from remote sensors or weather stations instead of from proximal data loggers, model outputs predicted that *Ae. albopictus* developmental rates were delayed and population growth rates were under-estimated (Figure 1) [60]. Thus, the environmental characteristics important to the survival of an IAVP vary considerably compared to those that can be measured or interpolated by currently available data [59,61], and the obliged use of sub-optimal proxy data may result in erroneous model outputs [48,61].

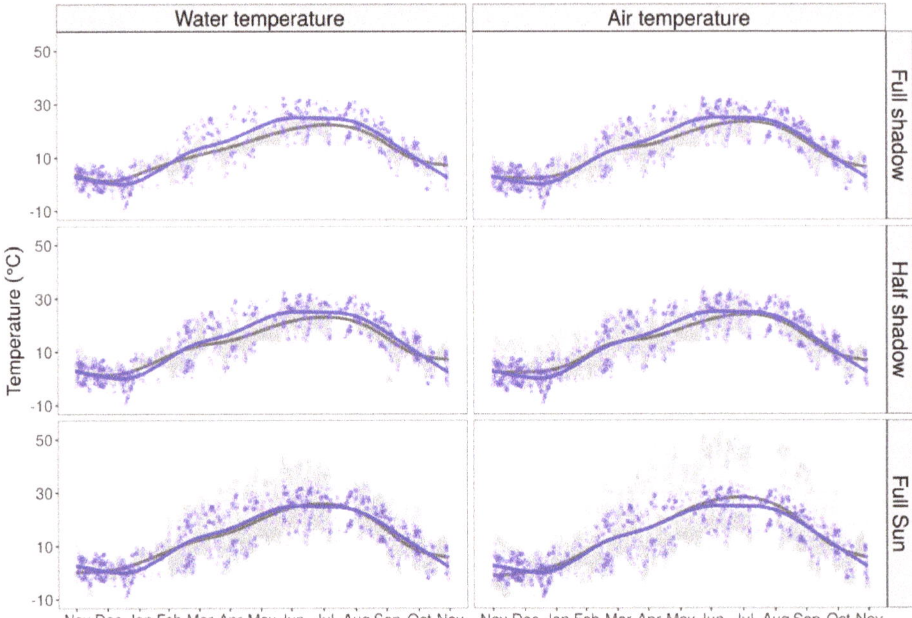

Figure 1. Temperature time series in an area invaded by *Aedes albopictus* mosquitoes in Italy (11.13°N, 46.2°E). Grey points and smoothed trends represent temperature measured at the microhabitat scale (i.e., an artificial, hard plastic, water filled container typically used for egg laying by this species) in different environmental settings: full-shadow, half-shadow, and full sun conditions. Temperatures were recorded inside (mosquito larvae habitat) and outside water (mosquito adult and egg habitat) using iButton® (Maxim Integrated, US) DS1923 data loggers at one hour intervals. Blue points and smoothed trends depict four-daily Land Surface Temperature (LST; MOD11A1 and MYD11A1 MODIS data) values, derived from the Moderate Resolution Imaging Sensor (MODIS) instruments, on-board the Terra and Aqua satellites. MODIS data were downloaded from a NASA server (https://lpdaac.usgs.gov/data_access), imported into GRASS GIS, and temperature values were extracted for each pixel (1 km resolution) where iButton sensors were placed (this figure was produced using data reported in [60]).

Strictly related to the low spatial resolution at which remotely sensed data are acquired, ecotones, i.e., where two macrohabitats intersect, for instance at the edge of a river, between mountains and valleys, green areas in a city or in catch-basins, are not currently well-captured by environmental data. However, ecotones can create microhabitat refugia in a macrohabitat that would otherwise be unsuitable. For example, Hoogstraal demonstrated that in the Nile Valley which is otherwise too dry for ticks, the soft tick *Ornithodoros sonrai* [Sautet and Witkowski, 1943] was able to colonize rodent burrows close to a permanent river, which provided adequate water and humidity [62]. Research has identified general patterns and mathematical relationships in the "buffering effect" of the physical structure of a microhabitat and has determined that, in general, within the microhabitat experienced by the tick, temperature and relative humidity are lower than that of the external environment typically measured for environmental data [63]. However, these patterns are influenced by a variety of factors, including the structure of the microhabitat and surrounding hydrography [63].

5. Attempting to Overcome the Lack of Microhabitat Data

Methods to capture, interpret, and produce remotely-sensed data that can be applied to SDMs are continually improving. In March 2019, Planet announced that they can provide satellite imagery from which NDVI can be derived at a resolution of 3–5 m, every 3 days (https://www.planet.com/pulse/developing-the-worlds-first-indicator-of-forest-carbon-stocks-emissions/). Additionally the project can gain NDVI at 0.8 m using Light Detecting and Ranging (LiDAR) sensors mounted on aircraft, but for only a single time point due to the costliness of this data collection process. Despite these improvements, there may be a considerable lag time between such data being available and being used in SDMs, which typically require time series data spanning multiple years to truly capture adequate information on the climate. In addition, high resolution data at such a large scale require intense computational power and expertise for use in SDMs, as high resolution satellite data brings with it challenges related to differentiating details between variables within the imagery, as well as new sources of noise [64]. For instance, a project attempted to use NDWI calculated from QuickBird satellite imagery at 2.44 m spatial resolution to locate potential habitat for invasive mosquitoes (e.g., swimming pools). However, ground truthing of the data showed that shadows cast upon swimming pools by surrounding trees or structures resulted in decreased NDWI values and reduced the ability to detect water bodies [65].

We understand that improving the quality of the remotely sensed data processing chain, including geometric and radiometric corrections, is a complex discipline in itself and takes time and an organized effort. However, we can take better ownership of the data that are currently available to us, and can follow the lead of other disciplines in doing this. A set of Essential Biodiversity Variables (EBV) have been identified to support biodiversity monitoring under the framework of the Group on Earth Observations Biodiversity Observation Network (GEO BON). Out of 21 candidate EBVs suggested by GEO BON, 14 EBVs have been identified as directly or indirectly measurable by remote sensing (https://geobon.org/ebvs/what-are-ebvs/) [66–68]. Two subsets of EBVs, focusing on Species Abundance (SA EBV) and Species Distribution (SD EBV), have been introduced and defined as a space-time-species-gram (cube), which can address species distribution or abundance irrespective of the taxonomy or scale [69]. This classification is facilitated by the availability of global, high-resolution, remotely-sensed data on environmental conditions and ecological species attributes. The framework has been optimized for biodiversity monitoring, but an equivalent product could be developed for relevant data pertaining to invasive species monitoring. Similarly, other areas of research have identified the need for environmental data that better meet the requirements of modelers, and have built high resolution and user friendly databases. For example, Bio-ORACLE (Ocean Rasters for Analysis of Climate and Environment) is a global dataset of environmental data which has been tailored for, and successfully implemented in, the distribution modeling of marine species [70–73]. Creating similar datasets that include environmental (both remotely and proximally sensed) data relevant to IAVP species at a fine spatial scale and a user friendly format could greatly improve the way in which currently available environmental data are used in IAVP SDMs. In addition, online data repositories, that include microhabitat data are available, such as DataONE (Data Observation Network for Earth, https://www.dataone.org/), JaLTER (Japan Long-Term Ecological Research Network; http://db.cger.nies.go.jp/JaLTER/metacat/style/skins/jalter-en/index.jsp) and the VLIZ: IMIS (The Flanders Marine Institute: Integrated Marine Information System, http://www.vliz.be/en/imis for example see [74]). Whilst these databases represent a great resource, people must be made aware that microclimatic data do exist, and centralization of microhabitat data in a well-structured repository could greatly facilitate data dissemination and utilization by the scientific community.

On a smaller scale, unmanned aerial vehicles (UAVs) and drones can be equipped with visible light, near-infrared, and/or thermal sensors to measure environmental variables, producing NDVI and surface temperature data at high resolution and at the desired scale [75,76]. UAVs have also been used to survey for bird and primate nesting and resting sites to estimate population numbers [77,78], and although this method is not currently suitable for the direct detection of IAVPs, host nesting sites, e.g., woodrat middens, or water bodies suitable for mosquito egg laying, could be surveyed using these techniques and used as a parameter for host availability in SDMs.

Environmental data at the microhabitat level can also be measured using data loggers; small sensors able to measure a range of variables paralleling those that can be remotely sensed, such as temperature, light, air velocity, barometric pressure, and relative humidity (e.g., see HOBO® U30 USB Station (U30-NRC) data logger; Bourne, MA). Many data loggers are small enough to be placed in almost any microhabitat, and can be programmed to record measurements at multiple intervals throughout a 24 h period. Data derived from such data loggers has been successfully used to model the extirpation and persistence of mammals (American pika, *Ochotona princeps* [Richardson, 1828]) [79,80] and thermal ecology potentially related to butterfly distribution (*Aglais urticae* Lepidoptera: Nymphalidae [Linnaeus, 1758], *Inachis io* Lepidoptera: Nymphalidae [Linnaeus, 1758] and *Polygonia c-album*, Lepidoptera: Nymphalidae [Linnaeus, 1758]) [81], and could no doubt also be applied to IAVP distributions. While a large number of data loggers need to be employed to collect sufficient data for species distribution modeling, requiring considerable resources to deploy and manage, these data could be supplemented by crowd-sourced means. Environmental data can now be collected from sensors within smart phones that can measure multiple variables, including temperature, pressure, and light, as well as from privately owned amateur weather stations and apps that ask citizens to report climatic data, such as amount of precipitation [82].

Despite the generalized application of coarse resolution data for modeling the distribution of IAVPs, overcoming the lack of data at a microhabitat scale has been attempted, although not necessarily for SDMs. An alternative and empirical strategy to improve our understanding of microhabitat thermal properties can be in the implementation of controlled experiments, that allow us to characterize microhabitat properties [83,84]. For example, studies have directly measured temperature within aquatic mosquito egg laying sites [48], and have recorded environmental variables in catch basins known to be egg laying sites for *Ae. albopictus* in Italy [58]. Both studies found that modeling mosquito population dynamics using these variables, rather than air temperature which is typically used, changes, and likely improves, the estimated development of the mosquito. An increasing number of scientific studies call for a better estimation of the thermal characteristics of mosquito microhabitats, specifically in order to achieve more reliable SDMs [48,58,85].

The strategies to overcome the lack of microhabitat data proposed above require intense use of resources, and we propose that microhabitat data collected by data loggers, crowd-sourcing, unmanned aerial vehicles, and/or controlled environmental studies are maintained in a database which can be freely contributed to, and be used by all those studying IAVP ecology. An open access database of microhabitat data could greatly facilitate the propagation of both collection and use of this type of environmental data.

6. Conclusions

Currently, environmental data that is used to inform ecological niche models largely relies on remotely sensed data, which is at a relatively course temporal and spatial resolution and does not accurately represent the microhabitat experienced by the species of interest, nor that at which activities informed by the prediction are executed. The predicted distribution of invasive arthropods resulting from models are therefore likely to be insufficient for direct application. The subsequent over- and/or under-estimations in IAVP distribution can have considerable consequences on control efforts, which may be informed by such predictions. We posit that consequently, efficiency and efficacy in the allocation of resources to control IAVPs are sub-optimal. The optimal resolution of environmental data relevant to IAVP ecology will likely vary according to the species under consideration, but we assume that this resolution will typically be <1 m and hourly. The scientific community may be far from having, for example, remotely-sensed measured temperature data at a 1 m spatial resolution or hourly temporal resolution for large extensions. However, we argue that any effort to improve the availability of data at a finer resolution than currently available will be of great benefit for modeling the distribution, abundance, or demographic rates of IAVP species. In the meantime we encourage modelers and ecologists to take a proactive approach in collecting fine resolution data using data loggers, crowd-sourcing, unmanned aerial vehicles and/or controlled environmental studies. We propose that these proximally-sensed data, as well as remotely-sensed data, be made open access in a user friendly database. We also hope that the suggestions made here for overcoming issues in environmental data for modeling IAVP distributions can be adapted and applied to species distribution modeling in other areas of research.

Author Contributions: Conceptualization, E.L.P. and M.M.; investigation, E.L.P., S.P., and M.M.; resources, E.L.P., S.P., D.R., and M.M.; formal analysis, M.M.; data curation, M.M.; Writing—Original Draft preparation, E.L.P., S.P., D.R., and M.M.; Writing—Review and Editing, E.L.P., S.P., and M.M.; visualization, E.L.P. and M.M.; supervision, M.M.; project administration, E.L.P.

Funding: E.L.P. and M.M. acknowledge funding support from the Pacific Southwest Regional Center of Excellence for Vector-Borne Diseases funded by the U.S. Centers for Disease Control and Prevention (Cooperative Agreement U01CK000516). M.M. also acknowledges funding from the National Aeronautics and Space Administration's Applied Sciences Program in Health and Air Quality (Grant NNX15AF36G).

Conflicts of Interest: The authors declare no conflict of interest.

References

1. Dukes, J.S.; Mooney, H.A. Does global change increase the success of biological invaders? *Trends Ecol. Evol.* **1999**, *14*, 135–139. [CrossRef]
2. Pyšek, P.; Richardson, D.M. Invasive species, environmental change and management, and health. *Annu. Rev. Environ. Resour.* **2010**, *35*, 25–55. [CrossRef]
3. Charles, H.; Dukes, J.S. Impacts of Invasive Species on Ecosystem Services. In *Biological Invasions*; Ecological Studies, Nentwig, W., Eds.; Springer: Berlin/Heidelberg, Germany, 2007; pp. 217–237, ISBN 978-3-540-36920-2.
4. U.S. Fish and Wildlife Services. The Cost of Invasive Species. 2012. Available online: https://www.fws.gov/verobeach/pythonpdf/costofinvasivesfactsheet.pdf (accessed on 20 February 2019).
5. Baly, A.; Toledo, M.E.; Lambert, I.; Benítez, E.; Rodriguez, K.; Rodriguez, E.; Vanlerberghe, V.; der Stuyft, P.V.; Baly, A.; Toledo, M.E.; et al. Cost of intensive routine control and incremental cost of insecticide-treated curtain deployment in a setting with low *Aedes aegypti* infestation. *Rev. Soc. Bras. Med. Trop.* **2016**, *49*, 418–424. [CrossRef]
6. Oreska, M.P.J.; Aldridge, D.C. Estimating the financial costs of freshwater invasive species in Great Britain: A standardized approach to invasive species costing. *Biol. Invasions* **2011**, *13*, 305–319. [CrossRef]
7. Grard, G.; Caron, M.; Mombo, I.M.; Nkoghe, D.; Mboui Ondo, S.; Jiolle, D.; Fontenille, D.; Paupy, C.; Leroy, E.M. Zika virus in Gabon (central Africa)—2007: A new threat from *Aedes albopictus*? *PLoS Negl. Trop. Dis.* **2014**, *8*, e2681. [CrossRef]
8. Rainey, T.; Occi, J.L.; Robbins, R.G.; Egizi, A. Discovery of *Haemaphysalis longicornis* (Ixodida: Ixodidae) parasitizing a sheep in New Jersey, United States. *J. Med. Entomol.* **2018**, *55*, 757–759. [CrossRef]
9. Robles, N.J.C.; Han, H.J.; Park, S.-J.; Choi, Y.K. Epidemiology of severe fever and thrombocytopenia syndrome virus infection and the need for therapeutics for the prevention. *Clin. Exp. Vaccine Res.* **2018**, *7*, 43–50. [CrossRef]
10. Sanders, C.J.; Mellor, P.S.; Wilson, A.J. Invasive arthropods. *Rev. Sci. Tech.* **2010**, *29*, 273–286. [CrossRef]
11. Simberloff, D. How much information on population biology is needed to manage introduced species? *Conserv. Biol.* **2003**, *17*, 83–92. [CrossRef]
12. Malanson, G.P.; Walsh, S.J. A Geographical Approach to Optimization of Response to Invasive Species. In *Science and Conservation in the Galapagos Islands: Frameworks & Perspectives*; Social and Ecological Interactions

in the Galapagos Islands, Walsh, S.J., Mena, C.F., Eds.; Springer: New York, NY, USA, 2013; pp. 199–215, ISBN 978-1-4614-5794-7.

13. Carter, G.A.; Lucas, K.L.; Blossom, G.A.; Lassitter, C.L.; Holiday, D.M.; Mooneyhan, D.S.; Fastring, D.R.; Holcombe, T.R.; Griffith, J.A. Remote sensing and mapping of tamarisk along the Colorado River, USA: A comparative use of summer-acquired hyperion, Thematic Mapper and QuickBird Data. *Remote Sens.* **2009**, *1*, 318–329. [CrossRef]

14. Rocchini, D.; Boyd, D.S.; Féret, J.-B.; Foody, G.M.; He, K.S.; Lausch, A.; Nagendra, H.; Wegmann, M.; Pettorelli, N. Satellite remote sensing to monitor species diversity: Potential and pitfalls. *Remote Sens. Ecol. Conserv.* **2015**, *2*, 25–36. [CrossRef]

15. Eklundh, L.; Johansson, T.; Solberg, S. Mapping insect defoliation in Scots pine with MODIS time-series data. *Remote Sens. Environ.* **2009**, *113*, 1566–1573. [CrossRef]

16. Jepsen, J.U.; Hagen, S.B.; Høgda, K.A.; Ims, R.A.; Karlsen, S.R.; Tømmervik, H.; Yoccoz, N.G. Monitoring the spatio-temporal dynamics of geometrid moth outbreaks in birch forest using MODIS-NDVI data. *Remote Sens. Environ.* **2009**, *113*, 1939–1947. [CrossRef]

17. GrrlScientist. Massive Swarm of Ladybugs Detected by California Weather Radar. Forbes. 2019. Available online: https://www.forbes.com/sites/grrlscientist/2019/06/10/160-square-mile-swarm-of-ladybugs-detected-by-california-weather-radar/#54ac52054d64 (accessed on 7 August 2019).

18. Nansen, C. The potential and prospects of proximal remote sensing of arthropod pests. *Pest Manag. Sci.* **2016**, *72*, 653–659. [CrossRef]

19. Rey, B.; Aleixos, N.; Cubero, S.; Blasco, J. Xf-Rovim. A field robot to detect olive trees infected by *Xylella fastidiosa* using proximal sensing. *Remote Sens.* **2019**, *11*, 221. [CrossRef]

20. Dunn, A.M.; Hatcher, M.J. Parasites and biological invasions: Parallels, interactions, and control. *Trends Parasitol.* **2015**, *31*, 189–199. [CrossRef]

21. Elith, J.; Graham, C.H. Do they? How do they? WHY do they differ? On finding reasons for differing performances of species distribution models. *Ecography* **2009**, *32*, 66–77. [CrossRef]

22. Shabani, F.; Kumar, L. Should species distribution models use only native or exotic records of existence or both? *Ecol. Inform.* **2015**, *29*, 57–65. [CrossRef]

23. Gutierrez, A.P.; Ponti, L.; Dalton, D.T. Analysis of the invasiveness of spotted wing Drosophila (*Drosophila suzukii*) in North America, Europe, and the Mediterranean Basin. *Biol. Invasions* **2016**, *18*, 3647–3663. [CrossRef]

24. Proestos, Y.; Christophides, G.K.; Ergüler, K.; Tanarhte, M.; Waldock, J.; Lelieveld, J. Present and future projections of habitat suitability of the Asian tiger mosquito, a vector of viral pathogens, from global climate simulation. *Philos. Trans. R. Soc. B* **2015**, *370*, 20130554. [CrossRef]

25. Lobo, J.M.; Jiménez-Valverde, A.; Hortal, J. The uncertain nature of absences and their importance in species distribution modelling. *Ecography* **2010**, *33*, 103–114. [CrossRef]

26. Cord, A.F.; Meentemeyer, R.K.; Leitão, P.J.; Václavík, T. Modelling species distributions with remote sensing data: Bridging disciplinary perspectives. *J. Biogeogr.* **2013**, *40*, 2226–2227. [CrossRef]

27. Hay, S.I. An overview of remote sensing and geodesy for epidemiology and public health application. *Adv. Parasitol.* **2000**, *47*, 1–35.

28. Kyba, C.C.M.; Garz, S.; Kuechly, H.; De Miguel, A.S.; Zamorano, J.; Fischer, J.; Hölker, F. High-resolution imagery of Earth at night: New sources, opportunities and challenges. *Remote Sens.* **2015**, *7*, 1–23. [CrossRef]

29. Elvidge, C.; Baugh, K.; Hobson, V.; Kihn, E.; Kroehl, H.; Davis, E.; Cocero, D. Satellite inventory of human settlements using nocturnal radiation emissions: A contribution for the global toolchest. *Glob. Chang. Biol.* **1997**, *3*, 387–395. [CrossRef]

30. Marcantonio, M.; Rizzoli, A.; Metz, M.; Rosà, R.; Marini, G.; Chadwick, E.; Neteler, M. Identifying the environmental conditions favouring West Nile virus outbreaks in Europe. *PLoS ONE* **2015**, *10*, e0121158. [CrossRef]

31. Pareeth, S.; Salmaso, N.; Adrian, R.; Neteler, M. Homogenised daily lake surface water temperature data generated from multiple satellite sensors: A long-term case study of a large sub-Alpine lake. *Sci. Rep.* **2016**, *6*, 31251. [CrossRef]

32. Metz, M.; Andreo, V.; Neteler, M. A new fully gap-free time series of land surface temperature from MODIS LST data. *Remote Sens.* **2017**, *9*, 1333. [CrossRef]

33. European Environment Agency Copernicus Land Monitoring Service—Corine Land Cover. Available online: https://www.eea.europa.eu/data-and-maps/data/copernicus-land-monitoring-service-corine (accessed on 20 August 2019).

34. Land Monitoring Service CORINE Land Cover—Copernicus Land Monitoring Service. Available online: https://land.copernicus.eu/pan-european/corine-land-cover (accessed on 20 August 2019).

35. Yang, L.; Jin, S.; Danielson, P.; Homer, C.; Gass, L.; Bender, S.M.; Case, A.; Costello, C.; Dewitz, J.; Fry, J.; et al. A new generation of the United States National Land Cover Database: Requirements, research priorities, design, and implementation strategies. *ISPRS J. Photogramm. Remote Sens.* **2018**, *146*, 108–123. [CrossRef]

36. Friedl, M.A.; Sulla-Menashe, D.; Tan, B.; Schneider, A.; Ramankutty, N.; Sibley, A.; Huang, X. MODIS Collection 5 global land cover: Algorithm refinements and characterization of new datasets. *Remote Sens. Environ.* **2010**, *114*, 168–182. [CrossRef]

37. Fick, S.E.; Hijmans, R.J. WorldClim 2: New 1-km spatial resolution climate surfaces for global land areas. *Int. J. Climatol.* **2017**, *37*, 4302–4315. [CrossRef]

38. Hijmans, R.J.; Cameron, S.E.; Parra, J.L.; Jones, P.G.; Jarvis, A. Very high resolution interpolated climate surfaces for global land areas. *Int. J. Climatol.* **2005**, *25*, 1965–1978. [CrossRef]

39. Jarnevich, C.S.; Reynolds, L.V. Challenges of predicting the potential distribution of a slow-spreading invader: A habitat suitability map for an invasive riparian tree. *Biol. Invasions* **2011**, *13*, 153–163. [CrossRef]

40. Kreakie, B.J.; Fan, Y.; Keitt, T.H. Enhanced migratory waterfowl distribution modeling by inclusion of depth to water table data. *PLoS ONE* **2012**, *7*, e30142. [CrossRef]

41. Pascoe, E.L.; Marcantonio, M.; Caminade, C.; Foley, J.E. Modeling potential habitat for *Amblyomma* tick species in California. *Insects* **2019**, *10*, 201. [CrossRef]

42. Cornes, R.C.; van der Schrier, G.; van den Besselaar, E.J.M.; Jones, P.D. An ensemble version of the E-OBS temperature and precipitation data sets. *J. Geophys. Res. Atmos.* **2018**, *123*, 9391–9409. [CrossRef]

43. Bedia, J.; Herrera, S.; Gutiérrez, J.M. Dangers of using global bioclimatic datasets for ecological niche modeling. Limitations for future climate projections. *Glob. Planet. Chang.* **2013**, *107*, 1–12. [CrossRef]

44. Hofstra, N.; Haylock, M.; New, M.; Jones, P.; Frei, C. Comparison of six methods for the interpolation of daily, European climate data. *J. Geophys. Res. Atmos.* **2008**, *113*, D21. [CrossRef]

45. Center for Diseases Control and Prevention (CDC). *ESTIMATED Potential Range of Aedes aegypti and Aedes albopictus in the United States, 2017*; CDC: Atlanta, GA, USA, 2018.

46. Rose, R.I. Pesticides and public health: Integrated methods of mosquito management. *Emerg. Infect. Dis.* **2001**, *7*, 17–23. [CrossRef] [PubMed]

47. Born, W.; Rauschmayer, F.; Bräuer, I. Economic evaluation of biological invasions—A survey. *Ecol. Econ.* **2005**, *55*, 321–336. [CrossRef]

48. Paaijmans, K.P.; Imbahale, S.S.; Thomas, M.B.; Takken, W. Relevant microclimate for determining the development rate of malaria mosquitoes and possible implications of climate change. *Malar. J.* **2010**, *9*, 196. [CrossRef] [PubMed]

49. Schulze, T.L.; Jordan, R.A.; Hung, R.W. Effects of microscale habitat physiognomy on the focal distribution of *Ixodes scapularis* and *Amblyomma americanum* (Acari: Ixodidae) nymphs. *Environ. Entomol.* **2002**, *31*, 1085–1090. [CrossRef]

50. Beck-Johnson, L.M.; Nelson, W.A.; Paaijmans, K.P.; Read, A.F.; Thomas, M.B.; Bjørnstad, O.N. The importance of temperature fluctuations in understanding mosquito population dynamics and malaria risk. *R. Soc. Open Sci.* **2017**, *4*, 160969. [CrossRef] [PubMed]

51. Gray, J.S.; Estrada-Peña, A.; Vial, L. Ecology of nidicolous ticks. In *Biology of Ticks*; Oxford University Press: Oxford, UK, 2014; Volume 2, pp. 39–60.

52. Burks, C.S.; Stewart, R.L.; Needham, G.R.; Lee, R.E. The role of direct chilling injury and inoculative freezing in cold tolerance of *Amblyomma americanum*, *Dermacentor variabilis* and *Ixodes scapularis*. *Physiol. Entomol.* **1996**, *21*, 44–50. [CrossRef]

53. Yoder, J.A.; Tank, J.L. Similarity in critical transition temperature for ticks adapted for different environments: Studies on the water balance of unfed ixodid larvae. *Int. J. Acarol.* **2006**, *32*, 323–329. [CrossRef]

54. Allingham, P.G. Effect of temperature on late immature stages of *Culicoides brevitarsis* (Diptera: Ceratopogonidae). *J. Med. Entomol.* **1991**, *28*, 878–881. [CrossRef] [PubMed]

55. Zellweger, F.; De Frenne, P.; Lenoir, J.; Rocchini, D.; Coomes, D. Advances in microclimate ecology arising from remote sensing. *Trends Ecol. Evol.* **2019**, *34*, 327–341. [CrossRef] [PubMed]

56. Chabot-Couture, G.; Nigmatulina, K.; Eckhoff, P. An environmental data set for vector-borne disease modeling and epidemiology. *PLoS ONE* **2014**, *9*, e94741. [CrossRef] [PubMed]

57. Vancutsem, C.; Ceccato, P.; Dinku, T.; Connor, S.J. Evaluation of MODIS land surface temperature data to estimate air temperature in different ecosystems over Africa. *Remote Sens. Environ.* **2010**, *114*, 449–465. [CrossRef]

58. Vallorani, R.; Angelini, P.; Bellini, R.; Carrieri, M.; Crisci, A.; Zeo, S.M.; Messeri, G.; Venturelli, C. Temperature characterization of different urban microhabitats of *Aedes albopictus* (Diptera Culicidae) in Central–Northern Italy. *Environ. Entomol.* **2015**, *44*, 1182–1192. [CrossRef]

59. Becker, N. *Mosquitoes and Their Control*; Springer: Berlin/Heidelberg, Germany; London, UK, 2010; ISBN 978-3-540-92874-4.

60. Marcantonio, M. Environmental Modelling and Spatial Ecology with Focus on Invasive *Aedes* Mosquitoes and Emergent Mosquito-Borne Pathogens. Ph.D. Thesis, Technische Universität Berlin, Berlin, Germany, 2017.

61. Yee, D.A. Thirty years of *Aedes albopictus* (Diptera: Culicidae) in America: An introduction to current perspectives and future challenges. *J. Med. Entomol.* **2016**, *53*, 989–991. [CrossRef]

62. Hoogstraal, H. A Preliminary, Annotated list of ticks (Ixodoidea) of the Anglo-Egyptian Sudan. *J. Parasitol.* **1954**, *40*, 304. [CrossRef]

63. Vial, L. Biological and ecological characteristics of soft ticks (Ixodida: Argasidae) and their impact for predicting tick and associated disease distribution. *Parasite* **2009**, *16*, 191–202. [CrossRef] [PubMed]

64. Nagendra, H.; Rocchini, D. High resolution satellite imagery for tropical biodiversity studies: The devil is in the detail. *Biodivers. Conserv.* **2008**, *17*, 3431. [CrossRef]

65. McFeeters, S.K. Using the normalized difference water index (NDWI) within a geographic information system to detect swimming pools for mosquito abatement: A practical approach. *Remote Sens.* **2013**, *5*, 3544–3561. [CrossRef]

66. O'Connor, B.; Secades, C.; Penner, J.; Sonnenschein, R.; Skidmore, A.; Burgess, N.D.; Hutton, J.M. Earth observation as a tool for tracking progress towards the Aichi Biodiversity Targets. *Remote Sens. Ecol. Conserv.* **2015**, *1*, 19–28. [CrossRef]

67. Schmeller, D.S.; Weatherdon, L.V.; Loyau, A.; Bondeau, A.; Brotons, L.; Brummitt, N.; Geijzendorffer, I.R.; Haase, P.; Kuemmerlen, M.; Martin, C.S.; et al. A suite of essential biodiversity variables for detecting critical biodiversity change. *Biol. Rev.* **2018**, *93*, 55–71. [CrossRef] [PubMed]

68. Vihervaara, P.; Auvinen, A.-P.; Mononen, L.; Törmä, M.; Ahlroth, P.; Anttila, S.; Böttcher, K.; Forsius, M.; Heino, J.; Heliölä, J.; et al. How Essential Biodiversity Variables and remote sensing can help national biodiversity monitoring. *Glob. Ecol. Conserv.* **2017**, *10*, 43–59. [CrossRef]

69. Jetz, W.; McGeoch, M.A.; Guralnick, R.; Ferrier, S.; Beck, J.; Costello, M.J.; Fernandez, M.; Geller, G.N.; Keil, P.; Merow, C.; et al. Essential biodiversity variables for mapping and monitoring species populations. *Nat. Ecol. Evol.* **2019**, *3*, 539. [CrossRef] [PubMed]

70. Bridge, T.; Beaman, R.; Done, T.; Webster, J. Predicting the location and spatial extent of submerged coral reef habitat in the Great Barrier Reef World Heritage Area, Australia. *PLoS ONE* **2012**, *7*, e48203. [CrossRef] [PubMed]

71. Jueterbock, A.; Tyberghein, L.; Verbruggen, H.; Coyer, J.A.; Olsen, J.L.; Hoarau, G. Climate change impact on seaweed meadow distribution in the North Atlantic rocky intertidal. *Ecol. Evol.* **2013**, *3*, 1356–1373. [CrossRef] [PubMed]

72. Quillfeldt, P.; Masello, J.F.; Navarro, J.; Phillips, R.A. Year-round distribution suggests spatial segregation of two small petrel species in the South Atlantic. *J. Biogeogr.* **2013**, *40*, 430–441. [CrossRef]

73. Tyberghein, L.; Verbruggen, H.; Pauly, K.; Troupin, C.; Mineur, F.; De Clerck, O. Bio-ORACLE: A global environmental dataset for marine species distribution modelling: Bio-ORACLE marine environmental data rasters. *Glob. Ecol. Biogeogr.* **2012**, *21*, 272–281. [CrossRef]

74. Guden, R.M.E.; Vafeiadou, A.M.; De Meester, N.; Derycke, S.; Moens, T. *Relative Abundance Data of 4 Cryptic Lineages of the Nematode Litoditis Marina in a Saltmarsh Habitat in the Western-Scheldt Estuary*; The Flanders Marine Institute: Integrated Marine Information System: London, UK, 2018.

75. Díaz-Delgado, R.; Ónodi, G.; Kröel-Dulay, G.; Kertész, M. Enhancement of ecological field experimental research by means of UAV multispectral sensing. *Drones* **2019**, *3*, 7. [CrossRef]

76. Harvey, M.C.; Rowland, J.V.; Luketina, K.M. Drone with thermal infrared camera provides high resolution georeferenced imagery of the Waikite geothermal area, New Zealand. *J. Volcanol. Geotherm. Res.* **2016**, *325*, 61–69. [CrossRef]

77. Afán, I.; Máñez, M.; Díaz-Delgado, R. Drone monitoring of breeding waterbird populations: The case of the Glossy ibis. *Drones* **2018**, *2*, 42. [CrossRef]

78. Bonnin, N.; Van Andel, A.C.; Kerby, J.T.; Piel, A.K.; Pintea, L.; Wich, S.A. Assessment of chimpanzee nest detectability in drone-acquired images. *Drones* **2018**, *2*, 17. [CrossRef]

79. Wilkening, J.L.; Ray, C.; Beever, E.A.; Brussard, P.F. Modeling contemporary range retraction in Great Basin pikas (*Ochotona princeps*) using data on microclimate and microhabitat. *Quat. Int.* **2011**, *235*, 77–88. [CrossRef]

80. Beever, E.A.; Ray, C.; Mote, P.W.; Wilkening, J.L. Testing alternative models of climate-mediated extirpations. *Ecol. Appl.* **2010**, *20*, 164–178. [CrossRef]

81. Bryant, S.R.; Thomas, C.D.; Bale, J.S. The influence of thermal ecology on the distribution of three nymphalid butterflies. *J. Appl. Ecol.* **2002**, *39*, 43–55. [CrossRef]

82. Muller, C.L.; Chapman, L.; Johnston, S.; Kidd, C.; Illingworth, S.; Foody, G.; Overeem, A.; Leigh, R.R. Crowdsourcing for climate and atmospheric sciences: Current status and future potential. *Int. J. Climatol.* **2015**, *35*, 3185–3203. [CrossRef]

83. Thomas, S.M.; Obermayr, U.; Fischer, D.; Kreyling, J.; Beierkuhnlein, C. Low-temperature threshold for egg survival of a post-diapause and non-diapause European aedine strain, *Aedes albopictus* (Diptera: Culicidae). *Parasites Vectors* **2012**, *5*, 100. [CrossRef] [PubMed]

84. Watt, J.H.; van den Berg, S. Chapter 15. Semi-Controlled Environments: Field Research. In *Research Methods for Communication Science*; Allyn & Bacon: Boston, MA, USA, 2002; pp. 227–241.

85. Asare, E.O.; Tompkins, A.M.; Amekudzi, L.K.; Ermert, V. A breeding site model for regional, dynamical malaria simulations evaluated using in situ temporary ponds observations. *Geospat. Health* **2016**, *11*, 390. [CrossRef] [PubMed]

Data Descriptor

System for Collecting, Processing, Visualization, and Storage of the MT-Monitoring Data

Elena Bataleva, Anatoly Rybin and Vitalii Matiukov *

Research Station of the Russian Academy of Sciences in Bishkek, Bishkek 720049, Kyrgyzstan
* Correspondence: vitaliy_cowboy@mail.ru

Received: 20 May 2019; Accepted: 12 July 2019; Published: 14 July 2019

Abstract: On the basis of the Research Station of the Russian Academy of Sciences in Bishkek, a unique scientific infrastructure—a complex geophysical station—is successfully functioning, realizing a monitoring of geodynamic processes, which includes research on the network of points of seismological, geodesic, and electromagnetic observations on the territory of the Bishkek Geodynamic Proving Ground located in the seismically active zone of the Northern Tien Shan. The scientific and practical importance of monitoring the geodynamical activity of the Earth's crust takes place not only in seismically active regions, but also in the areas of the location of particularly important objects, mining, and hazardous industries. Therefore, it seems highly relevant to create new software and hardware to study geodynamic processes in the earth's crust of seismically active zones, based on integrated monitoring of the geological environment in the widest possible depth range. The use of modern information technology in such studies provides an effective data management tool. The considering system for collecting, processing, and storing monitoring electromagnetic data of the Bishkek geodynamic proving ground can help overcome the scarcity of experimental data in the field of Earth sciences.

Dataset: For general use, a center for collective use of scientific equipment "Integrated geodynamic research" (CCU IGR) was created, on the basis of the Research Station of the Russian Academy of Sciences in Bishkek (RS RAS) (http://ckp-rf.ru/auth/). Through it, you can register and get access to data that is laid out for general use. In open access on the Internet, EDI-files on the MANAS profile are posted at http://ds.iris.edu/spud/emtf.

Dataset License: CC-BY

Keywords: database; geophysical monitoring; magnetotelluric monitoring; processing

---

1. Summary

The complex of regional geophysical works, including magnetotelluric studies, is carried out in almost all major tectonic provinces worldwide. One of the tasks of such a complex is to study the geodynamic state of the regions and assess the development and distribution of hazardous geological processes. Tien Shan region is one of the most tectonically active. This paper discusses the information aspects of the developed technology of multidisciplinary geophysical monitoring of geodynamic processes in the Earth's crust seismically active regions. The approach to the created technology is based on the integrated use of structural-functional and object-oriented information models. The developed structural-functional information model describes the processes of obtaining, storing and converting raw electromagnetic data, measured by magnetotelluric soundings method (MTS), and the object-oriented model used for describing the data itself (initial, intermediate, and final) and the relationships between them. The models are built using CASE tools All Fusion (Business Process Modeler—BPwin) and Power Designer, to define the boundaries and hierarchical structure of the developed system. The created technology provides an effective information system for integrated geophysical monitoring of geodynamic processes originating in the earth's crust of the seismically active zone of the Northern Tien Shan (the territory of the Bishkek geodynamic polygon) [1–3]. The main element of the complex geophysical monitoring is electromagnetic observations with a natural source of electromagnetic fields, which include magnetotelluric (MT) continuous observations of changes in the electrical parameters of the geoelectric cross-section at stationary points, continuous geomagnetic observations of the full vector T of the geomagnetic field at stationary points of the network and periodic observations at controlled points served by mobile stations. MT observations are used to determine variations of electromagnetic parameters in the Tien Shan lithosphere to a depths of 100 km and to study their relationship with geodynamic processes, occurring at these depths.

This work represents the results of research related to the development of azimuthal magnetotelluric monitoring techniques, which consists of analyzing the obtained time series of electromagnetic parameters in order to determine the contribution of each of the components of the impedance tensor to the informativeness of monitoring studies [3]. On the basis of the correlation analysis of gravitational tidal effects and the results of magnetotelluric monitoring, an additional test is carried out, the previously identified azimuthal dependence of the environmental stress sensitivity. When performing modern monitoring studies, scientists have to face an unprecedented amount of data that is subject to orderly storage, processing, graphical visualization and analysis [4]. Both stations are located on the territory of the Bishkek geodynamic proving ground, which in turn is part of the Northern Tien Shan seismic zone (Figure 1), and data is recorded around the clock in the period of 0.01–1000 s. Over the years of research, a catalog of geoelectrical data based on magnetotelluric soundings (MTS) and magnetovariational soundings (MVS) made in a series of regional and local profiles in the range of periods from 0.06 to 1800 s, created in the Tien Shan region, has been constantly updated. The catalog also includes the results of deep magnetotelluric soundings (periods up to 10,000 s). Information characterizing the parameters of the network of magnetotelluric observations (observation points and their coordinates) is contained in the catalog of the regional network of MTS, MVS cost center. To date, the established regional network of MTS and MVS for stationary observation points and profiles covers almost the entire territory of Tien Shan, within Kyrgyzstan and the surrounding areas.

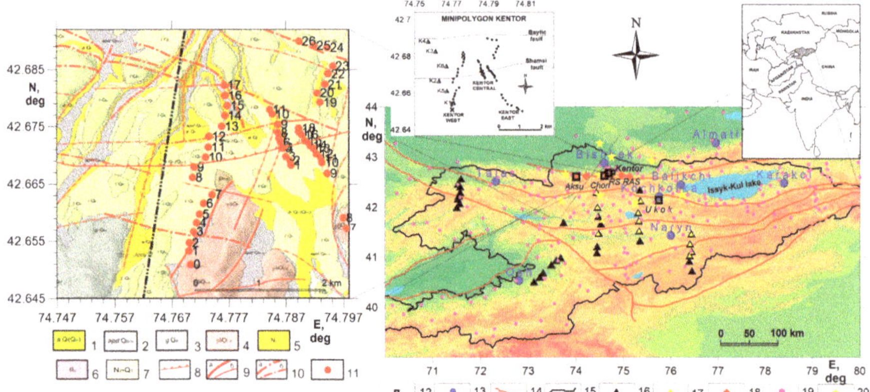

Figure 1. Location map of points of the geophysical monitoring, performed on the territory of the Central Tien Shan: 1—modern alluvial boulder-pebble deposits; 2—alluvial boulder-pebble deposits of the first above-flood terrace; 3—blocky-pebble glacial deposits; 4—Early and Middle Ordovician granodiorite; 5—Miocene pebble-crumbly-sandy strata; 6—Riphean complexes (undivided); 7—Pliocene–Pleistocene alluvial boulder deposits of the Sharpyldak series; 8—Shamsi fault; 9—activated faults and fracture zones of the basement: a—main; b—secondary; 10—activated faults assumed under the cover of modern sediments: a—main; b—secondary; 11—points of regime magnetotelluric soundings (MTS); 12—points of stationary and profile magnetotelluric observations; 13—settlements; 14—main fault structures; 15—the border of Kyrgyzstan; 16—points of regime deep MTS; 17—points of regime deep MTS of 2018; 18—points of electromagnetic monitoring; 19—points of the network of GPS observations; 20—KNET teleseismic network sites.

The data acquisition and information processing system of magnetotelluric monitoring allows collecting and accumulating data from a variety of monitoring observation points—stationary, regime, and profile (Figure 1). Monitoring was performed to study geodynamic processes in the Earth's crust and upper mantle based on the calculation of the transfer functions between the components of the magnetotelluric field with high temporal resolution in order to study their temporal dynamics. The final result of such monitoring, from a formal point of view, is a set of time series of various data [5,6]. In the practice of monitoring geodynamical processes, statistical methods of data analysis are widely used. In particular, a correlation analysis is used to determine the degree of the interrelation of the observed data series. Time series are formed, which are used to study changes in the recorded parameters over time and to isolate anomalies associated with the preparation of strong earthquakes [7,8]. Programs are designed for visualization, processing, and analysis of time series. They have a convenient user interface. They implemented arithmetic, statistical, and other functions for working with time series. It is possible to edit drawings (graphs) on the screen, save, and print them.

2. Data Description

According to the results of continuous monitoring of electromagnetic, geomagnetic, GPS, gravimetric, and seismic observations, banks of primary data of the territory of the Bishkek geodynamic proving ground are formed and a catalog of earthquakes is compiled. As an example, consider the procedure for collecting data from magnetotelluric monitoring.

2.1. Data Collection Procedure

The monitoring network continuously records the MT field on the embedded flash memory of the Phoenix MTU-5D instrumentation. The duration of the recording depends on the amount of flash memory and registration parameters. The registration parameters indicate the polling frequency and the duration of the recording. At the maximum polling frequency, the recording time is about 20 days, after which data is copied from the flash memory to a laptop, and the equipment is serviced and restarted. When working on the MTS profile, the measurement mode depends on the objectives of the study and is seasonal. In an ordinary observation point on a profile, the duration of an MT-field recording is a time interval from several hours to several days, which is determined by the depth of soundings. To check the performance of the station, a test recording of about an hour is made. The most informative is the night registration.

2.2. Structure and Data Processing of the Magnetotelluric Sounding (MTS) Method

The primary time series files of magnetotelluric data are stored in two binary files one of which saves the data of high and middle-frequency band (2400 and 150 Hz) at intervals of a few seconds from the beginning of the minute, while the second file continuously saves low-frequency data (24 Hz).

Time series are accompanied by a small binary file which saves registration parameters. To process time series data, use the SSMT 2000 program from the standard set of Phoenix software. As a result of this processing, average daily MT monitoring records are obtained stored as binary files. Work with these files is performed using the GSPlot (General Spectra Plot) program from the standard set of Phoenix software. The GSPlot program allows to visually view the transformants of the MT data and also presents them in a table form.

The data storage scheme of magnetotelluric soundings processing is based on the data storage scheme in the international data exchange standard MT. In this standard, sensing data at a point is written to a file with the extension EDI (Electrical Data Interchange). EDI files are obtained using the MT-Correct processing application program developed by the North-West geophysical company and saved in ASCII format, in contrast to the primary binary files.

2.3. Structure and Storage of Magnetotelluric (MT) Data

All MT sounding material, both source material and processing results, are placed in archives on working computers, in a database and in an external archive on CDs. In the MT database, the material is classified by year of observation, by profiles and measurement points.

The data of the MT-monitoring are located in the directories corresponding to the names of the items—Aksu-mon and Chon-mon, in which folders with the number of years of monitoring are created.

The data on the MT-profiles are in the directories with the name of the profile and year of work.

3. Methods

The geophysical monitoring database (DB) of the RS RAS includes an electromagnetic observation database with an artificial source of electromagnetic field, electromagnetic observations with a natural source of electromagnetic field, geodetic GPS observations on a local network and geomagnetic observations and can be considered as a single distributed database (Distributed DataBase—DDB) [9] of geophysical monitoring. These databases play the role of local databases located in different nodes of the corporate and/or global computer network. DDB, as defined by Data [10], can be considered as a loosely coupled network structure whose nodes are local databases.

Local databases are autonomous, independent, and self-defined; access to them is provided through the DBMS. Connections between nodes are replicated data streams. DDB topology is a star structure.

For organizing the collection, storage of data, and processing of the results of MT monitoring, the As IS model was developed. The model was developed in the BPWin [11] environment in the form of data flow diagrams (DFD-diagrams) and is presented in Figure 2.

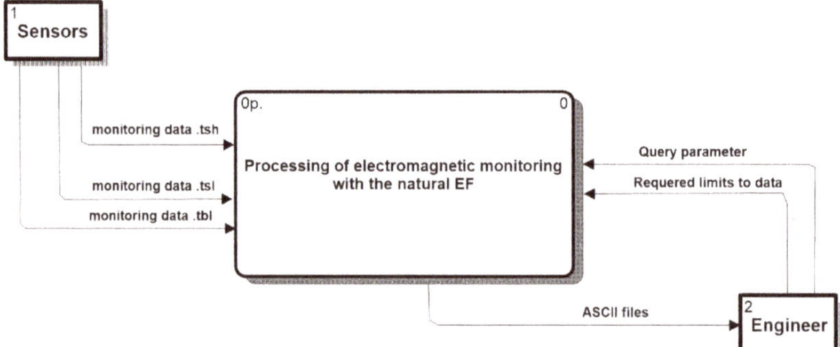

Figure 2. Data streams in magnetotelluric (MT)-monitoring (Model As IS).

For on-line access to the results of MT monitoring, a model of a distributed interactive system of access to the results of magnetotelluric monitoring in the form of a data flow diagram (DFD diagrams) was developed. This model is essentially an As To Be model and is presented in Figure 3.

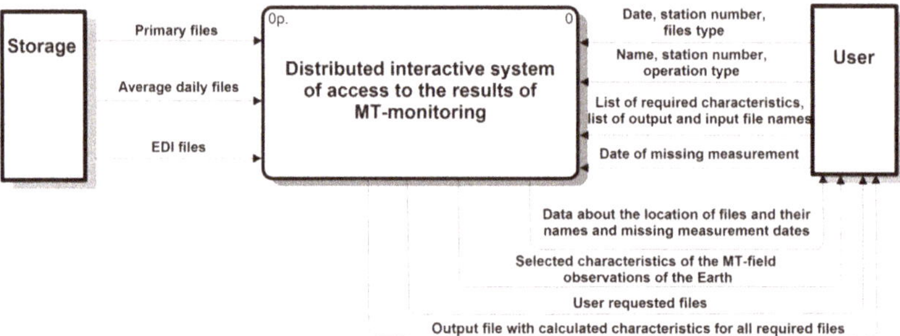

Figure 3. Model of a distributed interactive system of access to the result of magnetotelluric (MT)-monitoring (Model As To Be).

On the system model, the main process is allocated: distributed interactive system of access to the results of monitoring MT and two external entities: User and Storage. The repository is a storage medium on the external medium with respect to the system, on which the primary Phoenix station files, the average daily (processed) files, and EDI files are stored. Figure 4 shows an example of the MT data outputs that was obtained from the database.

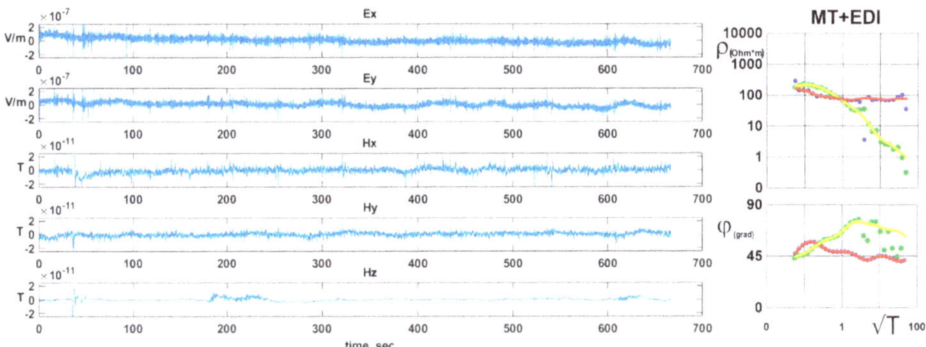

Figure 4. The example of the magnetotelluric (MT) data outputs that was obtained from database. The left side shows the time series source data—5 components of the electromagnetic field. In the right side sample processed data. Blue and green dots show processed curves that are obtained in binary format, red and yellow solid lines—smooth curves in EDI format.

The detailing of the main process is carried out in the form of a 1-level DFD, including 3 main subprocesses (Processing of primary files, Processing of daily average files, Processing of EDI files) and 2 auxiliary subprocesses (Construction of correlation diagrams and MT time-frequency monitoring), interacting with the Server.

Description of the Main Functions of the System

Based on the models discussed above, a logical database structure was developed in the ERwin environment [10]. Visual Basic .Net 2008 and SQL Server 2000 DBMS are selected as programming tools. The developed distributed interactive system of access to the results of magnetotelluric monitoring has the following functionalities.

1. Creating a database. To create a database of MT monitoring files, you must enter the name of the server on which the database will be created and the name of the future database, as well as select the type of authentication for the server on the local computer or to create a database on the remote computer.

2. Setting up the software system and filling the database. At this point, you can select or enter: the name of the server to which you want to connect, and the name of the database created by the program, in which the data about the MT monitoring files will be stored. In addition, you can choose the path to the files, up to folders with stations.

3. Database update.

4. Search for files by date and coordinates. The file search is possible by date, based on the type of the files you are looking for, the time period in which the necessary files are located, the station number from which the files were received. The search by coordinates implies the search for files by the latitude and longitude of the location of the stations.

5. Copying files.

6. Processing MT monitoring data.

7. Construction of time-frequency series of MT monitoring.

8. Construction of correlation diagrams.

4. User Notes

Usage: for collecting and processing geophysical information, in particular for measuring, recording, and processing the electrical and magnetic components of the natural electromagnetic field, in the study of geodynamic processes occurring in the Earth's crust and upper mantle using electrical survey methods. Thus, the developed software system makes it possible to increase the efficiency of processing MT monitoring data by significantly reducing the time spent searching for the necessary information, the ability to quickly view the newly received data, and create a distributed database of monitoring MT observations over a period of about 15 years. Currently, the system is in the trial operation of the Research Station of the Russian Academy of Sciences.

5. Patents

As part of these studies, the database "Local database of magnetotelluric data in the system of geophysical monitoring of geodynamic processes in the Earth's crust of seismically active regions" was registered. Certificate No. 2012621291, issued on 07.12.2012. Copyright holder FGBUN Research Station of the Russian Academy of Sciences in Bishkek. Authors: Rybin A.K., Matiukov V.E., Desyatkov G.A., Lychenko N.M., Manzhikova S.T. At the present, this database is actively used and developed.

Author Contributions: Conceptualization, E.B.; Data curation, A.R.; Investigation, A.R.; Methodology, E.B.; Project administration, A.R.; Software, V.M.; Visualization, V.M.; Writing—original draft, E.B.; Writing—review & editing, V.M.

Funding: The presented researches are carried out within the fulfilment of the State Assignment by the Research Station of the Russian Academy of Sciences (subject AAAA-A19-119020190063-2 (0155-2019-0001)) and with the financial support of the Russian Foundation for Basic Research (Project 17-05-00654).

Conflicts of Interest: The authors declare no conflict of interest.

References

1. Rybin, A.; Fox, L.; Ingerov, O.; Schelochkov, G.; Safronov, I.; Batalev, V.; Bataleva, E. MT monitoring experiment in the Northern Tien Shan seismogenic zone/CD ROM: Abstracts. In Proceedings of the 10-th Scientific Assembly of the International Association of Geomagnetism and Aeronomy, Toulouse, France, 18–25 July 2005.
2. Rybin, A.K.; Batalev, V.Y.; Bataleva, E.A.; Matiukov, V.E.; Desyatkov, G.A.; Lychenko, N.M.; Manzhikova, S.T.; Ten, V. Development of a distributed interactive system of access to the results of magnetotelluric monitoring. In Proceedings of the Problems of geodynamics and geoecology of intracontinental orogens VII International Symposium, Bishkek city, Kyrgyz Republic, 19–24 June 2011; pp. 179–187.
3. Bataleva, E.A.; Batalev, V.Y.; Rybin, A.K. On the question of the interrelation between variations in crustal electrical conductivity and geodynamical processes. *Izvestiya Phys. Solid Earth* **2013**, *49*, 402–410. [CrossRef]
4. Lyubushin, A.A. Multidimensional analysis of the time series of geophysical monitoring systems. *Izvestiya Phys. Solid Earth* **1993**, *3*, 103–108.
5. Bataleva, E.A.; Batalev, V.Y. Development of Programs to Analyze the Data on Azimuthal Magnetotelluric Monitoring Part 1. Analysis of magnetotelluric monitoring data. *Vestnik Kyrgyz-Slavic University* **2014**, *14*, 3–7.
6. Bataleva, E.A.; Batalev, V.Y. Development of Programs to Analyze the Data on Azimuthal Magnetotelluric Monitoring Part 2. Development of software for analysis of MT monitoring data. *Vestnik Kyrgyz-Slavic University* **2014**, *14*, 8–12.
7. Bataleva, E.A.; Zabinyakova, O.B.; Batalev, V.Y. Software Development for Monitoring Electromagnetic Parameters of the Bishkek Geodynamic Providing Group. *Vestnik Kyrgyz-Slavic University* **2017**, *17*, 144–149.
8. Bataleva, E.A.; Zabinyakova, O.B.; Batalev, V.Y. Development for Magnetotelluric Profile Monitoring Cantor's Minipoligon. *Vestnik Kyrgyz-Slavic University* **2017**, *17*, 150–153.
9. Date, C.J. What is distributed database? Available online: ftp://sohoftp.nascom.nasa.gov/pub/www/publications/chapter10.pdf (accessed on 3 March 2019).

10. Infological modeling. Available online: http://www.rus-lib.ru/book/28/ps/01/027-044.html (accessed on 15 April 2019).
11. Maklakov, S.V. BPwin and ERwin—CASE-development tools for information systems. M.: Dialog-MEPI. 1999. Available online: https://studfiles.net/preview/942930/page:10/ (accessed on 10 May 2019).

MDPI

St. Alban-Anlage 66

4052 Basel

Switzerland

Tel. +41 61 683 77 34

Fax +41 61 302 89 18

www.mdpi.com

Data Editorial Office

E-mail: data@mdpi.com

www.mdpi.com/journal/data